T0230439

Computational Synthesis and Creative Systems

Series editors

François Pachet, Paris, France
Pablo Gervás, Madrid, Spain
Andrea Passerini, Trento, Italy
Mirko Degli Esposti, Bologna, Italy

Creativity has become the motto of the modern world: everyone, every institution, and every company is exhorted to create, to innovate, to think out of the box. This calls for the design of a new class of technology, aimed at assisting humans in tasks that are deemed creative.

Developing a machine capable of synthesizing completely novel instances from a certain domain of interest is a formidable challenge for computer science, with potentially ground-breaking applications in fields such as biotechnology, design, and art. Creativity and originality are major requirements, as is the ability to interact with humans in a virtuous loop of recommendation and feedback. The problem calls for an interdisciplinary perspective, combining fields such as machine learning, artificial intelligence, engineering, design, and experimental psychology. Related questions and challenges include the design of systems that effectively explore large instance spaces; evaluating automatic generation systems, notably in creative domains; designing systems that foster creativity in humans; formalizing (aspects of) the notions of creativity and originality; designing productive collaboration scenarios between humans and machines for creative tasks; and understanding the dynamics of creative collective systems.

This book series intends to publish monographs, textbooks and edited books with a strong technical content, and focuses on approaches to computational synthesis that contribute not only to specific problem areas, but more generally introduce new problems, new data, or new well-defined challenges to computer science.

More information about this series at http://www.springer.com/series/15219

Noor Shaker · Julian Togelius
Mark J. Nelson

Procedural Content
Generation in Games

 Springer

Noor Shaker
Department of Architecture, Design
  and Media Technology
Aalborg University Copenhagen
  (AAU CPH)
Copenhagen
Denmark

Mark J. Nelson
The MetaMakers Institute
Falmouth University
Penryn, Cornwall
UK

Julian Togelius
Department of Computer Science
  and Engineering
New York University
Brooklyn, NY
USA

ISSN 2509-6575   ISSN 2509-6583   (electronic)
Computational Synthesis and Creative Systems
ISBN 978-3-319-82643-1   ISBN 978-3-319-42716-4   (eBook)
DOI 10.1007/978-3-319-42716-4

Printed on acid-free paper

This Springer imprint is published by Springer Nature
The registered company is Springer International Publishing AG
The registered company address is: Gewerbestrasse 11, 6330 Cham, Switzerland

Preface

Welcome to the Procedural Content Generation in Games book.[1] This is, as far as we know, the first textbook about procedural content generation in games. As far as we know it is also the first book-length overview of the research field. We hope you find it useful, whether you are studying in a course, on your own, or are a researcher.

We wrote this book for two reasons. The first reason was that all three of us were doing research on PCG in games, and we wanted a good overview. As we come from somewhat different methodological backgrounds, we realized that many researchers did not know about methods that had been developed in other communities. For example, researchers using logic programming and those using evolutionary computation might not know that the other type of algorithms was applicable to the same problem; and researchers coming from computer graphics might not even know that artificial intelligence methods are being used for PCG problems. As PCG in games has just recently started to be seen as its own research field, this was not surprising, but pointed to the need for a book such as this one.

The second reason was that we were teaching a course on PCG (in fact, entitled simply "Procedural Content Generation in Games") at the IT University of Copenhagen, where at the time the three of us were faculty members. When this course was started in 2010, it was probably the first of its kind in the world. Naturally, there was no textbook to teach it from, so we assembled a syllabus out of academic papers, mostly recent ones. As we taught the course in subsequent years, the syllabus matured, and we felt that we were ready to turn the content of our lectures into a textbook.

In writing the book, we based it on the structure of the existing course. In fact the first draft of this textbook was written quite literally as part of the fourth iteration of the course in autumn 2013. A draft of each chapter was completed in advance of the corresponding lecture, and given out as a handout to accompany the lecture. This ensured that a complete draft of the textbook was written within one semester, and perhaps more importantly ensures that the book is designed to be used as a textbook. Unfortunately, adding polish and finalising each chapter took a lot longer,

---

[1] Throughout the book, we will often use PCG as an acronym for procedural content generation.

which explains why the book did not come out in 2014. We believe however that the added time to work on the book was worth it, as the final product is much better than those early drafts were.

As you will see, the book is not strictly divided according to either methods or application domains. Most chapters introduce both a domain and method. This follows how we structured our PCG course, which we did in order to make the course more engaging and easier to teach: new methods are introduced together with interesting and relevant domains that demonstrate why they are practically useful, and can be used as settings for lab exercises to further experiment with the methods.

We decided early on that we wanted to involve many of the most active people in the research field, in particular those who had written the papers we relied on when teaching the course initially. Therefore, most chapters are coauthored with other researchers. This ensures that we have the most relevant expertise for the topic of each chapter.

As mentioned above, one of the main purposes of this book is as a textbook. Its origins as a set of course notes has also helped ensure that the book is "battle-tested" and ready to teach from. In particular, the book can be used as the main course text for a graduate-level or advanced undergraduate-level course on Procedural Content Generation in Games. It is assumed that students are familiar with basic artificial intelligence concepts (in particular heuristic search and basic concepts of logic and machine learning) and it is very beneficial (though not strictly necessary) that students have some experience of game development and using a game engine.

This book could be used as course literature in several ways. One is to base each lecture on a specific chapter, and assign a few recent papers from the literature related to that chapter as additional reading for that week. The assignment at the end of each chapter could then be used as that week's assignment, and a conventional pen-and-paper exam could be held at the end of the course. Another way of organizing such a course, more closely aligning with the way the original course at the IT University of Copenhagen is taught, could be to use the first half of the semester for intensive lectures, covering two chapters per week. The second half of the semester is then used for group projects.[2] Finally, you can always use parts of the book, for example if you want to teach PCG as part of a larger course in AI for games. Most chapters are reasonably self-contained, with the most important dependencies being on the first two chapters, which establish core concepts and terminology. Therefore it is advisable to start with chapters 1 and 2 even if only using parts of the book.

The book is accompanied by a webpage, pcgbook.com, which contains digital versions of the book chapters, along with example lecture slides, links to relevant mailing groups and conferences, and other supplemental material. We welcome suggestions for new supplemental material (e.g. your own lecture slides) to add to the website. Our updated contact information can also be found there.

Of course, in any book-sized effort, one relies on the help of a large number of other people. Our first and foremost thanks go to our collaborators and co-authors on the various chapters of the book. We are also very grateful to our students in the

---

[2] Many good papers came out of the group projects from that course.

PCG course who endured various draft versions of the book, as well as our actual lectures, and in many cases provided very useful feedback. Several other colleagues have provided useful feedback or helped out in other ways; the list includes Steve Dahlskog, Amy Hoover and Aaron Isaksen.

Noor Shaker, Julian Togelius, and Mark Nelson

Contents

List of Contributors

The authors of this book are Noor Shaker, Julian Togelius, and Mark Nelson, who have had overall responsibility for the book and contributed to all chapters. In addition, 14 domain experts have contributed to individual chapters. The full list of authors, in order of appearance as chapter authors, is below.

Noor Shaker
Department of Architecture, Design and Media Technology, Aalborg University Copenhagen, A. C. Meyers Vaenge 15, 2450 Copenhagen, Denmark, e-mail: ns@create.aau.dk

Julian Togelius
Department of Computer Science and Engineering, New York University, 2 MetroTech Center, Brooklyn, NY 11201, United States, e-mail: julian@togelius.com

Mark J. Nelson
The MetaMakers Institute, Falmouth University, Treliever Road, Penryn Cornwall TR10 9FE, United Kingdom, e-mail: mjn@anadrome.org

Antonios Liapis
Institute of Digital Games, University of Malta, Msida, MSD2080, Malta, e-mail: antonios.liapis@um.edu.mt

Ricardo Lopes
Department of Intelligent Systems, University of Delft, Mekelweg 4, 2628 CD, Delft, Netherlands, e-mail: r.lopes@tudelft.nl

Rafael Bidarra
Department of Intelligent Systems, University of Delft, Mekelweg 4, 2628 CD, Delft, Netherlands, e-mail: r.bidarra@tudelft.nl

Joris Dormans
Ludomotion, Reitzstraat 105,1091 ZA, Amsterdam, Netherlands, e-mail:
joris@ludomotion.com

Cameron Browne
Science and Engineering Faculty, Queensland University of Technology, 2 George
St, Brisbane City QLD 4000, Australia, e-mail: c.browne@qut.edu.au

Michael Cook
The MetaMakers Institute, Falmouth University, Treliever Road, Penryn Cornwall
TR10 9FE, United Kingdom, e-mail: mike@gamesbyangelina.org

Yun-Gyung Cheong
College of Information & Communication Engineering, Sungkyunkwan University,
25-2 Sungkyunkwan-ro, Jongno-gu, Seoul, Korea, e-mail: aimecca@skku.edu

Mark O. Riedl
College of Computing, Georgia Institute of Technology, North Avenue, Atlanta,
GA 30332, United States, e-mail: riedl@cc.gatech.edu

Byung-Chull Bae
School of Games, Hongik University, 94, Wausan-ro, Mapo-gu, Seoul, 04066,
Korea, e-mail: byuc@hongik.ac.kr

Adam M. Smith
Department of Computational Media, University of California Santa Cruz, 1156
High Street, Santa Cruz, CA 95064, United States, e-mail: adam@adamsmith.as

Daniel Ashlock
Department of Mathematics and Statistics, University of Guelph, 50 Stone Road
East, Guelph, Ontario, N1G 2W1, Canada, e-mail: dashlock@uoguelph.ca

Sebastian Risi
IT University of Copenhagen, Rued Langgaards Vej 7, 2300 Copenhagen S,
Denmark, e-mail: sebr@itu.dk

Georgios N. Yannakakis
Institute of Digital Games, University of Malta, Msida, MSD2080, Malta, e-mail:
georgios.yannakakis@um.edu.mt

Gillian Smith
College of Arts, Media and Design, Northeastern University, 360 Huntington
Avenue, Boston, Massachusetts 02115, United States, e-mail: gillian@ccs.neu.edu

Chapter 1
Introduction

Julian Togelius, Noor Shaker, and Mark J. Nelson

Abstract This chapter introduces the field of procedural content generation (PCG), as well as the book. We start by defining key terms, such as game content and procedural generation. We then give examples of games that use PCG, outline desirable properties, and provide a taxonomy of different types of PCG. Applications of and approaches to PCG can be described in many different ways, and another section is devoted to seeing PCG through the lens of design metaphors. The chapter finishes by providing an overview of the rest of the book.

1.1 What is procedural content generation?

You have just started reading a book about Procedural Content Generation in Games. This book will contain quite a lot of algorithms and other technical content, and plenty of discussion of game design. But before we get to the meat of the book, let us start with something a bit more dry: definitions. In particular, let us define Procedural Content Generation, which we will frequently abbreviate as PCG. The definition we will use is that *PCG is the algorithmic creation of game content with limited or indirect user input* [32]. In other words, PCG refers to computer software that can create game content on its own, or together with one or many human players or designers.

A key term here is "content". In our definition, content is most of what is contained in a game: levels, maps, game rules, textures, stories, items, quests, music, weapons, vehicles, characters, etc. The game engine itself is not considered to be content in our definition. Further, non-player character behaviour—NPC AI—is not considered to be content either. The reason for this narrowing of the definition of content is that within the field of artificial and computational intelligence in games, there is much more research done in applying CI and AI methods to character behaviour than there is on procedural content generation. While the field of PCG is mostly based on AI methods, we want to set it apart from the more "mainstream"

© Springer International Publishing Switzerland 2016
N. Shaker et al., *Procedural Content Generation in Games*, Computational
Synthesis and Creative Systems, DOI 10.1007/978-3-319-42716-4_1

use of game-based tasks to test AI algorithms, where AI is most often used to learn to play a game. Like all definitions (except perhaps those in mathematics), our definition of PCG is somewhat arbitrary and rather fuzzy around the edges. We will treat it as such, and are mindful that other people define the term differently. In particular, some would rather use the term "generative methods" for a superset of what we call PCG [8].

Another important term is "games". Games are famously hard to define (see Wittgenstein's discussion of the matter [36]), and we will not attempt this here. Suffice it to say that by games we mean such things as videogames, computer games, board games, card games, puzzles, etc. It is important that the content generation system takes the design, affordances and constraints of the game that it is being generated for into account. This sets PCG apart from such endeavours as generative art and many types of computer graphics, which do not take the particular constraints and affordances of game design into account. In particular, a key requirement of generated content is that it must be playable—it should be possible to finish a generated level, ascend a generated staircase, use a generated weapon or win a generated game.

The terms "procedural" and "generation" imply that we are dealing with computer procedures, or algorithms, that create something. A PCG method can be run by a computer (perhaps with human help), and will output something. A PCG *system* refers to a system that incorporates a PCG method as one of its parts, for example an adaptive game or an AI-assisted game design tool. This book will contain plenty of discussion of algorithms and quite a lot of pseudocode, and most of the exercises that accompany the chapters will involve programming.

To make this discussion more concrete, we will list a few things we consider to be PCG:

- A software tool that creates dungeons for an action adventure game such as *The Legend of Zelda* without any human input—each time the tool is run, a new level is created;
- a system that creates new weapons in a space shooter game in response to what the collective of players do, so that the weapons that a player is presented with are evolved versions of weapons other players found fun to use;
- a program that generates complete, playable and balanced board games on its own, perhaps using some existing board games as starting points;
- game engine middleware that rapidly populates a game world with vegetation;
- a graphical design tool that lets a user design maps for a strategy game, while continuously evaluating the designed map for its gameplay properties and suggesting improvements to the map to make it better balanced and more interesting.

In the upcoming chapters, you will find descriptions of all of those things described above. Let us now list a few things that we do not consider to be PCG:

- A map editor for a strategy game that simply lets the user place and remove items, without taking any initiative or doing any generation on its own;
- an artificial player for a board game;
- a game engine capable of integrating automatically generated vegetation.

Several other authors have tackled the issue of surveying PCG or part of the field we call PCG, though the overlap is far from complete [12, 25].

1.2 Why use procedural content generation?

Now that we know what PCG is, let us discuss the reasons for using and developing such methods. It turns out there are a number of different reasons.

Perhaps the most obvious reason to generate content is that it removes the need to have a human designer or artist generate that content. Humans are expensive and slow, and it seems we need more and more of them all the time. Ever since computer games were invented, the number of person-months that go into the development of a successful commercial game has increased more or less constantly.[1] It is now common for a game to be developed by hundreds of people over a period of a year or more. This leads to a situation where fewer games are profitable, and fewer developers can afford to develop a game, leading in turn to less risk-taking and less diversity in the games marketplace. Many of the costly employees necessary in this process are designers and artists rather than programmers. A game development company that could replace some of the artists and designers with algorithms would have a competitive advantage, as games could be produced faster and cheaper while preserving quality. (This argument was made forcefully by legendary game designer Will Wright in his talk "The Future of Content" at the 2005 Game Developers Conference, a talk which helped reinvigorate interest in procedural content generation.)

Of course, threatening to put them out their jobs is no way to sell PCG to designers and artists. We could therefore turn the argument around: content generation, especially embedded in intelligent design tools, can augment the creativity of individual human creators. This could make it possible for small teams without the resources of large companies, and even for hobbyists, to create content-rich games by freeing them from worrying about details and drudge work while retaining overall directorship of the games.

Both of these arguments assume that what we want to make is something like the games we have today. But PCG methods could also enable completely new types of games. To begin with, if we have software that can generate game content at the speed it is being "consumed" (played), there is in principle no reason why games need to end. For everyone who has ever been disappointed by their favourite game not having any more levels to clear, characters to meet, areas to explore, etc., this is an exciting prospect.

Even more excitingly, the newly generated content can be tailored to the tastes and needs of the player playing the game. By combining PCG with player modelling, for example through measuring and using neural networks to model the response of players to individual game elements, we can create player-adaptive games

---

[1] At least, this is true for "AAA" games, which are boxed games sold at full price worldwide. The recent rise of mobile games seems to have made single-person development feasible again, though average development costs are rising on that front too.

that seek to maximise the enjoyment of players. The same techniques could be used to maximise the learning effects of a serious game, or perhaps the addictiveness of a "casual" game.

Another reason for using PCG is that it might help us to be more creative. Humans, even those of the "creative" vein, tend to imitate each other and themselves. Algorithmic approaches might come up with radically different content than a human would create, through offering an unexpected but valid solution to a given content generation problem. Outside of games, this is a well-known phenomenon in e.g. evolutionary design.

Finally, a completely different but no less important reason for developing PCG methods is to understand design. Computer scientists are fond of saying that you don't really understand a process until you have implemented it in code (and the program runs). Creating software that can competently generate game content could help us understand the process by which we "manually" generate the content, and clarify the affordances and constraints of the design problem we are addressing. This is an iterative process, whereby better PCG methods can lead to better understanding of the design process, which in turn can lead to better PCG algorithms.

1.3 Games that use PCG

Overcoming the storage limitations of computers was one of the main driving forces behind the development of PCG techniques. The limited capabilities of home computers in the early 1980s constrained the space available to store game content, forcing designers to pursue other methods for generating and saving content. *Elite* [4] is one of the early games that solved this problem by storing the seed numbers used to procedurally generate eight galaxies each with 256 planets each with unique properties. Another classical example of the early use of PCG is the early-1980s game *Rogue*, a dungeon-crawling game in which levels are randomly generated every time a new game starts. Automatic generation of game content, however, often comes with tradeoffs; roguelike games can automatically generate compelling experiences, but most of them (such as *Dwarf Fortress* [1]) lack visual appeal.

Procedural content generation has received increasing attention in commercial games. *Diablo* [2] is an action role-playing hack-and-slash videogame featuring procedural generation for creating the maps, and the type, number and placement of items and monsters. PCG is a central feature in *Spore* [15] where the designs the players create are animated using procedural animation techniques. These personalised creatures are then used to populate a procedurally generated galaxy. *Civilization IV* [10] is a turn-based strategy game that allows unique gameplay experience by generating random maps. *Minecraft* [19] is a massively popular game featuring extensive use of PCG techniques to generate the whole world and its content. *Spelunky* [39, 38] is another notable 2D platform roguelike indie game that utilizes PCG to automatically generate variations of game levels (Figure 1.1). *Tiny Wings*

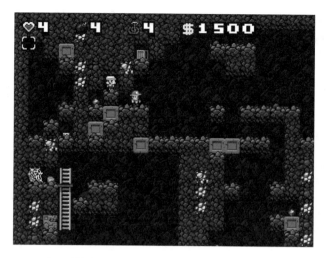

Fig. 1.1: Screenshot from *Spelunky*

[13] is yet another example of a mobile 2D game featuring a procedural terrain and texture generation system giving the game a different look with each replay.

1.4 Visions for PCG

As we have seen, procedural content generation has been a part of some published games for three decades. In the past few years, there has also been a surge in academic research on PCG, where researchers from very different academic backgrounds have brought their perspectives and methods to bear on the problems of game content generation. This has resulted in a number of new methods, and variations and combinations of old methods, some of which are in need of further research and development before being useful in actual games. The chapters of this book will present many of the most significant contributions of recent years' research.

To guide the research being done, it is useful to have some visions of where we might be going; this is analogous to lists of "unsolved problems" in some research fields such as mathematics and physics. The authors of a recent survey paper defined three such visions for procedural content generation [31]. These are things that we cannot do with current technology, and might never be possible to achieve exactly as stated, but serve to point out limitations of the state of the art and by extension interesting problems to work on.

1. *Multi-level, multi-content PCG* refers to a content generator that, for a given game engine and set of game rules, would be able to generate all of the content for the game such that the content is of high quality and fits together perfectly.

For example, given the engine and ruleset for the popular computer role-playing game *Skyrim*, this imaginary software would generate backstory, quests, characters, items, weapons, vegetation, terrain, graphics, etc. in such a fashion that it all becomes a coherent, believable new world and an enjoyable game to play.

2. *PCG-based game design* refers to creating games that do not only rely on procedural content generation, but for which PCG is an absolutely central part of the gameplay, so that if you took the content generation part away there would not be anything recognisable left of the game. Some progress has been made towards this, notably in games such as *Galactic Arms Race* [11] and *Endless Web* [28], but these games are still based on established game genres and core parts of the games could function without PCG.

3. *Generating complete games* refers to a generator capable of generating not only content for a given game, but the game itself. This means the rules, reward structures and graphical representation as well as the levels, characters, etc. Some work has been done in this direction, mainly to generate rules for different kinds of games [33, 7, 20, 9], but the rules generated are so far rather simplistic.

Much of the work described in the upcoming chapters can be seen as making progress towards one or several of these visions, but, as you will see, there is much work to be done. At the same time, it is important to keep in mind that it is equally worthwhile to develop generators for more narrowly defined tasks.

1.5 Desirable properties of a PCG solution

We can think of implementations of PCG methods as *solutions* to *content generation problems*. A content generation problem might be to generate new grass with a low level of detail which does not look completely weird within 50 milliseconds. It might also be to generate a truly original idea for a game mechanic after days of computing time, or it might be to polish in-game items to a perfect sheen in a background thread as they are being edited by a designer. The desirable—or required—properties of a solution are different for each application. The only constant is that there are usually tradeoffs involved, e.g. between speed and quality, or expressivity/diversity and reliability. Here is a list of common desirable properties of PCG solutions:

- *Speed:* Requirements for speed vary wildly, from a maximum generation time of milliseconds to months, depending on (amongst other things) whether the content generation is done during gameplay or during development of the game.
- *Reliability:* Some generators shoot from the hip, whereas others are capable of guaranteeing that the content they generate satisfies some given quality criteria. This is more important for some types of content than others, for example a dungeon with no exit or entrance is a catastrophic failure, whereas a flower that looks a bit weird just looks a bit weird without this necessarily breaking the game.

- *Controllability:* There is frequently a need for content generators to be controllable in some sense, so that a human user or an algorithm (such as a player-adaptive mechanism) can specify some aspects of the content to be generated. There are many possible dimensions of control, e.g. one might ask for a smooth oblong rock, a car that can take sharp bends and has multiple colours, a level that induces a sense of mystery and rewards perfectionists, or a small ruleset where chance plays no part.
- *Expressivity and diversity:* There is often a need to generate a diverse set of content, to avoid the content looking like it's all minor variations on a tired theme. At an extreme of non-expressivity, consider a level "generator" that always outputs the same level but randomly changes the colour of a single stone in the middle of the level; at the other extreme, consider a "level" generator that assembles components completely randomly, yielding senseless and unplayable levels. Measuring expressivity is a non-trivial topic in its own right, and designing level generators that generate diverse content without compromising on quality is even less trivial.
- *Creativity and believability:* In most cases, we would like our content not to look like it has been designed by a procedural content generator. There is a number of ways in which generated content can look generated as opposed to human-created.

1.6 A taxonomy of PCG

With the variety of content generation problems and methods that are now available, it helps to have a structure that can highlight the differences and similarities between approaches. In the following, we introduce a revised version of the taxonomy of PCG that was originally presented by Togelius et al. [34]. It consists of a number of dimensions, where an individual method or solution should usually be thought of as lying somewhere on a continuum between the ends of that dimension.

1.6.1 Online versus offline

PCG techniques can be used to generate content online, as the player is playing the game, allowing the generation of endless variations, making the game infinitely replayable and opening the possibility of generating player-adapted content, or offline during the development of the game or before the start of a game session. The use of PCG for offline content generation is particularly useful when generating complex content such as environments and maps. An example of the use of online content generation can be found in the game *Left 4 Dead* [35], a recently released first-person shooter game that provides dynamic experience for each player by analysing player behaviour on the fly and altering the game state accordingly using PCG techniques [3].

NERO [30] is an example of the use of AI techniques to allow the players to evolve real-time tactics for a squad of virtual soldiers. *Forza Motorsport* [17] is a car racing game where the Non-Player Characters (NPCs) can be trained offline to imitate the player's driving style and can later be used to drive on behalf of the player. Another important use of offline content generation is the creation and sharing of content. Some games such as *LittleBigPlanet* [16] and *Spore* [15] provide a content editor (level editor in the case of *LittleBigPlanet* and the *Spore* Creature Creator) that allows the players to edit and upload complete creatures or levels to a central online server where they can be downloaded and used by other players.

1.6.2 Necessary versus optional

PCG can be used to generate necessary game content that is required for the completion of a level, or it can be used to generate auxiliary content that can be discarded or exchanged for other content. The main distinctive feature between necessary and optional content is that necessary content should always be correct while this condition does not hold for optional content. An example of optional content is the generation of different types of weapons in first-person shooter games or the auxiliary reward items in Super Mario Bros. [21]. Necessary content can be the main structure of the levels in Super Mario Bros., or the collection of certain items required to pass to the next level.

1.6.3 Degree and dimensions of control

The generation of content by PCG can be controlled in different ways. The use of a random seed is one way to gain control over the generation space; another way is to use a set of parameters that control the content generation along a number of dimensions. Random seeds were used when generating the world in *Minecraft* [19], which means the same world can be regenerated if the same seed is used [18]. A vector of content features was used in [24] to generate levels for *Infinite Mario Bros.* [22] that satisfy a set of feature specifications.

1.6.4 Generic versus adaptive

Generic content generation refers to the paradigm of PCG where content is generated without taking player behaviour into account, as opposed to adaptive, personalised or player-centred content generation where player interaction with the game

Fig. 1.2: Three example weapons created in the *Galactic Arms Race* game for different players. Adapted from [11]

is analysed and content is created based on a player's previous behaviour. Most commercial games tackle PCG in a generic way, while adaptive PCG has been receiving increasing attention in academia recently. A recent extensive review of PCG for player-adaptive games can be found in [37].

Left 4 Dead [35] is an example of the use of adaptive PCG in a commercial game where an algorithm is used to adjust the pacing of the game on the fly based on the player's *emotional intensity*. In this case, adaptive PCG is used to adjust the difficulty of the game in order to keep the player engaged [3]. Adaptive content generation can also be used with another motive such as the generation of more content of the kind the player seems to like. This approach was followed in the *Galactic Arms Race* [11] game where the weapons presented to the player are evolved based on her previous weapon use and preferences. Figure 1.2 presents examples of evolved weapons for different players.

1.6.5 Stochastic versus deterministic

Deterministic PCG allows the regeneration of the same content given the same starting point and method parameters as opposed to stochastic PCG where recreating the same content is usually not possible. The regeneration of the galaxies in *Elite* [4] is an example of the deterministic use of PCG.

1.6.6 Constructive versus generate-and-test

In constructive PCG, the content is generated in one pass, as commonly done in roguelike games. Generate-and-test PCG techniques, on the other hand, alternate generating and testing in a loop, repeating until a satisfactory solution is generated. *Yavalath* [5] is a two-player board game generated completely by a computer program using the generate-and-test paradigm [7].

1.6.7 Automatic generation versus mixed authorship

Until recently, PCG has allowed limited input from game designers, who usually tweak the algorithm parameters to control and guide content generation while the main purpose of PCG remains the generation of infinite variations of playable content [39, 7, 1, 2]. However, a new interesting paradigm, has emerged that focuses on incorporating designer and/or player input through the design process. In this mixed-initiative paradigm, a human designer or player cooperates with the algorithm to generate the desired content.

Tanagra [29] is an example of a system where the designer draws part of a 2D level and a constraint satisfaction algorithm is used to generate the missing parts while retaining playability. Another example is the SketchaWorld framework [26], an interactive procedural sketching system for creating landscapes and cityscapes where designers can manually edit and tune the generated results while the virtual world model is kept consistent. Ropossum [23] is yet another recent example of the use of PCG for completing unfinished designs, suggesting modifications, handling constraints and testing for playability for the 2D physics-based game Cut the Rope [40].

1.7 Metaphors for PCG

In the phrase "procedural content generation system", we have discussed what the words "procedural", "content", and "generation" mean. But what about the word system? A PCG system is the generic term for any piece of software that does PCG. But these systems do different things, are used in different ways, and have quite different relationships to the overall game-design process. Some PCG systems try to help a designer out with a small part of the design process. Others try to provide a new way of working with game content. Some are interactive; others aren't. Some aim to do fully autonomous, creative game design; others aim to automate routine or common aspects of design.

To break this broad term, PCG system, into more specific kinds of systems, Khaled et al. [14] proposed four metaphors for thinking about how PCG systems relate to the game-design process. Some PCG systems are tools: instruments that give designers enhanced capabilities, in the way that a programmer's development environment or an architect's CAD system do. Others define new kinds of materials, allowing a designer to work in a new medium, the way stone, clay, and laser installations are different materials for an artist. Some PCG systems are intended to be designers themselves, carrying out fully autonomous design of parts or even entire games, rather than assisting game designers. Finally, some systems are primarily domain experts, carrying with them extensive knowledge of game design that can be used to critique or improve designs. Many systems can be viewed through more than one of these lenses, though few will exhibit all of them equally.

PCG *tools*, like non-PCG design tools, aim to improve a designer's workflow, but PCG tools do it by adding a generative component. A common example is a PCG-enhanced level editor. The level-editing tools included with many game engines already improve the level-design process by providing specialised ways of editing and laying out levels, rather than the designer having to do level design in a more generic tool, or entirely in code. A PCG-enhanced level editor adds a generative component to the traditional passive level editor. The *Tanagra* [29] level editor generates levels that fit a theory of rhythmic patterns in platformer games, which the designer can modify and add more constraints to, followed by re-generation of the relevant portions. This back-and-forth pattern, alternating procedural content generation and human editing, is called *mixed-initiative* generation, and is covered in Chapter 11. Among the visions for PCG discussed earlier in this chapter, "multi-level multi-content PCG" can be seen as using a tool metaphor.

PCG systems can also create new generative *materials* that a game designer manipulates and sculpts to produce content. A popular commercial example is *SpeedTree*. In one sense it's a tool for designing trees to place as scenery in videogames. But the way it does this is by turning trees into an interactive generative material: the designer can click and drag them around, add and remove branches, etc., and they always look like a tree, because the trees are procedurally generated in real time as the designer manipulates them. The fractal landscapes discussed in Chapter 4 are also a kind of procedurally generative material, which a designer manipulates to produce their desired landscapes. For the PCG vision of "PCG-based game design", the appropriate metaphor is material.

A procedural content *designer* has less interaction with the human designer, and instead has ambitions of designing content all on its own. In the limit case, a PCG designer turns into a fully autonomous game generator that creates new games, usually in a specific genre. Work on automatic game design is still at an exploratory stage, but promising prototype systems exist [20, 33, 7, 9]. A key challenge for a lead designer is that it must design not only the content *in* a game, but the rules of the game itself. Chapter 6 looks at these systems that generate rules and game mechanics. The PCG vision of "generating complete games" relies on a designer metaphor.

A procedural *domain expert* is a slightly different kind of system, full of knowledge about games or players, and able to apply it to critique and modify content. Often it will apply that expertise by being part of a system that also serves as a tool or a designer. A domain expert may have purely formal knowledge of games, such as what makes a particular set of rules elegant [6]. Or it may have extensive knowledge of human players, being able to predict what people will do in a game, and what they will find challenging, fun, or boring. For a PCG-based educational game, the domain expert may have pedagogical knowledge. For example, the procedural level generation in the fraction-teaching game *Refraction* is constrained so that generated levels meet the system's pedagogical goals [27]. Chapter 10 discusses the experience-driven PCG approach, which builds PCG systems that are experts in player behaviour and reactions.

1.8 Outline of the book

This book is structured as a series of chapters, co-written by the main authors of the book and the leading experts on the topic of each chapter. Most chapters are organised so that they introduce both a family of methods (e.g. fractals or grammars) and an application domain (e.g. plants or dungeons). The method is typically introduced through an example in the application domain, and the chapter then also discusses how the same method could be used for other domains or how different methods could be used for that domain. This structure is partly motivated by the interdisciplinary nature of PCG research and practice, where the algorithms used come from numerous different fields (and thus rarely build on each other) and game design knowledge is vital in all cases. Each chapter ends with a summary and typically also with a proposed lab exercise.

In Chapter 2 we present the search-based approach to procedural content generation, which is very versatile and which has recently been used in a large number of academic research projects as well as some released games. In the search-based approach, evolutionary algorithms are used to search for good game content using principles from Darwinian evolution. The two main challenges when building a search-based content generator are the evaluation function, which evaluates candidate content artefacts, and the content representation, which defines the search space for the algorithm. While this chapter contains several examples of content generators based on artificial evolution, there are further such examples scattered in the upcoming chapters.

Chapter 3 discusses the specific example of creating dungeons for roguelike games, and similar levels based on navigating a mostly two-dimensional space—for example, levels for platform games or first-person shooters. A number of fast and constructive algorithms for generating such levels are described. Some of these algorithms come from the game development community and are widely used in roguelikes such as *Diablo*. Others, such as cellular automata, have their origin in physics. We also describe the Mario AI framework, a common testbed for level generation algorithms based on a clone of *Super Mario Bros*.

Chapter 4 describes several algorithms with a background in computer graphics research, namely simple fractal algorithms and other noise algorithms. These are commonly used to produce terrains and complete landscapes, as well as textures and features such as clouds. While these algorithms are fast and reliable, they lack some forms of controllability. Therefore two other approaches to generating landscapes are presented, one search-based and one based on collections of agents.

Chapter 5 is about grammars. Grammars, common to computer science and linguistics, prove to be very useful for creating many types of game content. The chapter starts with the example of creating lifelike plants, which is a very common form of PCG; in fact, hundreds of AAA games from recent years feature procedurally generated vegetation based on grammars. But grammars can also be used for e.g. level generation; the rest of the chapter details how to use grammars for generating levels and missions for Zelda-style action-adventure games, and how to evolve grammars that generate *Super Mario Bros*. levels.

While some of the application domains of the previous chapters may be seen as somewhat peripheral, Chapter 6 addresses the problems of generating the absolutely most central part of any game: its rules. We describe a number of different attempts at generating rules for games, from board games to card games and arcade games. Some of these attempts are constructive, but most of them are search-based in one way or another. The chapter also describes the Video Game Description Language, a way of encoding game rules for simple arcade games of the kind you would find in the early 1980s—one of the purposes of this language is to enable automatic generation of complete games.

Most games feature stories of some kind, either backstories or interactive stories that the player can affect; stories can be seen as content, so Chapter 7 is devoted to the generation of game stories. It turns out that almost all methods of story generation are based on planning algorithms; planning is a classic AI method originally developed for robot control and now widely used in various domains. The chapter also discusses how story generation can be combined with map generation, so that game maps are generated that fit with the generated story.

Chapter 8 is focused on a single method, namely Answer Set Programming (ASP). This is a form of logic programming plus constraint satisfaction: a content generation method plus conditions are specified in a language called AnsProlog, and a solver produces all configurations of content that are compatible with the specified conditions. While this might seem rather abstract and mathematical, it has recently been demonstrated that certain PCG problems can be easily stated in AnsProlog form, and the results of the solver interpreted as game content. This yields a highly efficient method for creating some form of game content, for example levels for puzzle-like games.

Chapter 9 returns to the topic of Chapter 2, search-based PCG, and dwells on the question of how to represent the game content. Representation is important as it defines the shape of the search space and the ways in which it can be explored. This chapter demonstrates how a wise choice of representation can alter the style of the generated content as well as enable more effective search for content that better satisfies the evaluation function. Examples include flowers represented as neural networks and level generators represented as collections of agents.

One of the motivations for PCG is that it can enable player-adaptive games. Chapter 10 describes a framework for adapting games to the player, namely that of experience-driven PCG. We describe different methods for creating models of player experience based on data collected from players.

A theme throughout much of the book is that the relationship between procedural content generation and human game designers can be quite varied. PCG can be used in a highly automated way, but it can also be used in close coupling with the designer's own design choices. Chapter 11 looks at this close coupling explicitly, considering mixed-initiative systems, in which a human designer and a procedural content generation system collaborate to produce content.

Finally, Chapter 12 discusses how the quality of a PCG solution can be evaluated once it has been implemented.

1.9 Summary

Procedural content generation (PCG) in games is the algorithmic creation of game content with limited or indirect user input. PCG methods are developed and used for a number of different reasons, including saving development time and costs, increasing replayability, allowing for adaptive games, assisting designers and studying creativity and game design. While PCG algorithms have been used in some commercial games since the early 1980s, they are typically either used in a peripheral role or their scope is highly limited; current research in academia is trying to push the boundaries of what can be generated and with what quality it can be generated. Ideally, a PCG solution should be fast, reliable, controllable, expressive and creative, but in practice there are certain tradeoffs that need to be made between these properties. PCG solutions can be classified according to a relatively extensive taxonomy, which might help to identify their strengths and weaknesses. Another lens through which to understand a PCG system is the metaphor according to which it is used; here we can differentiate between using a system as tool, material, designer or domain expert. PCG algorithms are drawn from a variety of different fields, and this methodological diversity is evident from the table of contents of this book.

References

1. Adams, T.: (2006). Dwarf Fortress, Bay 12 Games
2. Blizzard North: (1997). Diablo, Blizzard Entertainment, Ubisoft and Electronic Arts
3. Booth, M.: The AI systems of Left 4 Dead. In: Keynote, Fifth Artificial Intelligence and Interactive Digital Entertainment Conference (AIIDE) (2009)
4. Braben, D., Bell, I.: (1984). Elite, Acornsoft, Firebird and Imagineer
5. Browne, C.: Yavalath (2007). URL http://www.cameronius.com/games/yavalath/
6. Browne, C.: Elegance in game design. IEEE Transactions on Computational Intelligence and AI in Games **4**(3), 229 –240 (2012)
7. Browne, C., Maire, F.: Evolutionary game design. IEEE Transactions on Computational Intelligence and AI in Games **2**(1), 1–16 (2010)
8. Compton, K., Osborn, J.C., Mateas, M.: Generative methods. In: Proceedings of the 4th Workshop on Procedural Content Generation in Games (2013)
9. Cook, M., Colton, S.: Multi-faceted evolution of simple arcade games. In: Proceedings of the 7th IEEE Conference on Computational Intelligence and Games (CIG), pp. 289–296 (2011)
10. Firaxis Games: (2005). Civilization IV, 2K Games & Aspyr
11. Hastings, E.J., Guha, R., Stanley, K.: Evolving content in the Galactic Arms race video game. In: Proceedings of the 5th IEEE Conference on Computational Intelligence and Games (CIG), pp. 241–248 (2009)
12. Hendrikx, M., Meijer, S., van der Velden, J., Iosup, A.: Procedural content generation for games: a survey. Transactions on Multimedia Computing, Communications and Applications **9**(1), 1 (2013)
13. Illiger, A.: (2011). Tiny Wings, Andreas Illiger
14. Khaled, R., Nelson, M.J., Barr, P.: Design metaphors for procedural content generation in games. In: Proceedings of the 2013 ACM SIGCHI Conference on Human Factors in Computing Systems, pp. 1509–1518 (2013)
15. Maxis: (2008). Spore, Electronic Arts

16. Media Molecule, SCE Cambridge Studio, Tarsier Studios, Double Eleven, XDev, United Front Games: (2008). Little Big Planet, Sony Computer Entertainment Europe
17. Microsoft Game Studios: (2005). Forza Motorsport, Microsoft
18. Minecraft Wiki: Minecraft wiki. URL http://www.minecraftwiki.net/
19. Mojang: (2011). Minecraft, Mojang and Microsoft Studios
20. Nelson, M.J., Mateas, M.: Towards automated game design. In: AI*IA 2007: Artificial Intelligence and Human-Oriented Computing, pp. 626–637. Springer (2007). Lecture Notes in Computer Science 4733
21. Nintendo Creative Department: (1985). Super Mario Bros., Nintendo
22. Persson, M.: Infinite Mario Bros. URL http://www.mojang.com/notch/mario/
23. Shaker, M., Shaker, N., Togelius, J.: Ropossum: An authoring tool for designing, optimizing and solving Cut the Rope levels. In: Proceedings of the AAAI Conference on Artificial Intelligence and Interactive Digital Entertainment. AAAI Press, pp. 215 – 216 (2013)
24. Shaker, N., Togelius, J., Yannakakis, G.N.: Towards automatic personalized content generation for platform games. In: Proceedings of the AAAI Conference on Artificial Intelligence and Interactive Digital Entertainment (AIIDE), pp. 63–68 (2010)
25. Smelik, R., De Kraker, K., Tutenel, T., Bidarra, R., Groenewegen, S.: A survey of procedural methods for terrain modelling. In: Proceedings of the CASA Workshop on 3D Advanced Media In Gaming And Simulation (3AMIGAS) (2009)
26. Smelik, R., Tutenel, T., de Kraker, K., Bidarra, R.: Integrating procedural generation and manual editing of virtual worlds. In: Proceedings of the Workshop on Procedural Content Generation in Games. ACM (2010)
27. Smith, A.M., Andersen, E., Mateas, M., Popović, Z.: A case study of expressively constrainable level design automation tools for a puzzle game. In: Proceedings of the 7th International Conference on the Foundations of Digital Games, pp. 156–163 (2012)
28. Smith, G., Othenin-Girard, A., Whitehead, J., Wardrip-Fruin, N.: PCG-based game design: creating Endless Web. In: Proceedings of the International Conference on the Foundations of Digital Games, pp. 188–195. ACM (2012)
29. Smith, G., Whitehead, J., Mateas, M.: Tanagra: A mixed-initiative level design tool. In: Proceedings of the Fifth International Conference on the Foundations of Digital Games, pp. 209–216. ACM (2010)
30. Stanley, K., Bryant, B., Miikkulainen, R.: Real-time neuroevolution in the NERO video game. IEEE Transactions on Evolutionary Computation 9(6), 653–668 (2005)
31. Togelius, J., Champandard, A.J., Lanzi, P.L., Mateas, M., Paiva, A., Preuss, M., Stanley, K.O.: Procedural content generation: Goals, challenges and actionable steps. In: S.M. Lucas, M. Mateas, M. Preuss, P. Spronck, J. Togelius (eds.) Dagstuhl Seminar 12191: Artificial and Computational Intelligence in Games, pp. 61–75. Dagstuhl (2013)
32. Togelius, J., Kastbjerg, E., Schedl, D., Yannakakis, G.N.: What is procedural content generation?: Mario on the borderline. In: Proceedings of the 2nd Workshop on Procedural Content Generation in Games (2011)
33. Togelius, J., Schmidhuber, J.: An experiment in automatic game design. In: Proceedings of the 4th IEEE Symposium on Computational Intelligence and Games (CIG), pp. 111–118 (2008)
34. Togelius, J., Yannakakis, G.N., Stanley, K., Browne, C.: Search-based procedural content generation. Applications of Evolutionary Computation pp. 141–150 (2010)
35. Valve Corporation: (2008). Left 4 Dead, Valve Corporation
36. Wittgenstein, L.: Philosophical Investigations. Blackwell (1953)
37. Yannakakis, G.N., Togelius, J.: Experience-driven procedural content generation. IEEE Transactions on Affective Computing 2(3), 147–161 (2011)
38. Yu, D.: Spelunky. Boss Fight Books (2016)
39. Yu, D., Hull, A.: (2008). Spelunky, Independent
40. ZeptoLab: (2010). Cut the Rope

Chapter 2
The search-based approach

Julian Togelius and Noor Shaker

Abstract Search-based procedural content generation is the use of evolutionary computation and similar methods to generate game content. This chapter gives an overview of this approach to PCG, and lists a number of core considerations for developing a search-based PCG solution. In particular, we discuss how to best represent content so that the content space becomes searchable, and how to create an evaluation function that allows for effective search. Three longer examples of using search-based PCG to evolve content for specific games are given.

2.1 What is the search-based approach to procedural content generation?

There are many different approaches to generating content for games. In this chapter, we will introduce the *search-based approach*, which has been intensively investigated in academic PCG research in recent years. In search-based procedural content generation, an evolutionary algorithm or some other stochastic search/optimisation algorithm is used to search for content with the desired qualities. The basic metaphor is that of design as a search process: a good enough solution to the design problem exists within some space of solutions, and if we keep iterating and tweaking one or many possible solutions, keeping those changes which make the solution(s) better and discarding those that are harmful, we will eventually arrive at the desired solution. This metaphor has been used to describe the design process in many different disciplines: for example, Will Wright (designer of *SimCity* and *The Sims*) described the game design process as search in his talk at the 2005 Game Developers Conference [30]. Others have previously described the design process in general, and in other specialised domains such as architecture, the design process can be conceptualised as search and implemented as a computer program [29, 2].

The core components of the search-based approach to solving a content generation problem are the following:

N. Shaker et al., *Procedural Content Generation in Games*, Computational Synthesis and Creative Systems, DOI 10.1007/978-3-319-42716-4_2

- A *search algorithm*. This is the "engine" of a search-based method. As we will see, often relatively simple evolutionary algorithms work well enough, though sometimes there are substantial benefits to using more sophisticated algorithms that take e.g. constraints into account, or that are specialised for a particular content representation.
- A *content representation*. This is the representation of the artefacts you want to generate, e.g. levels, quests or winged kittens. The content representation could be anything from an array of real numbers to a graph to a string. The content representation defines (and thus also limits) what content can be generated, and determines whether effective search is possible.
- One or more *evaluation functions*. An evaluation function is a function from an artefact (an individual piece of content) to a number indicating the quality of the artefact. The output of an evaluation function could indicate e.g. the playability of a level, the intricacy of a quest or the aesthetic appeal of a winged kitten. Crafting an evaluation function that reliably measures the aspect of game quality that it is meant to measure is often among the hardest tasks in developing a search-based PCG method.

This chapter will describe each of these components in turn. It will also discuss several examples of search-based methods for generating different types of content for different types of games.

2.2 Evolutionary search algorithms

An evolutionary algorithm is a stochastic search algorithm loosely inspired by Darwinian evolution through natural selection. The core idea is to keep a *population* of *individuals* (also called chromosomes or candidate solutions), which in each *generation* are evaluated, and the *fittest* (highest evaluated) individuals get the chance to *reproduce* and the least fit are removed from the population. A generation can thus be seen as divided into *selection* and reproduction phases. In your backyard, a generation of newly born rabbits may be subject to selection by the hungry wolf who eats the slowest of the litter, with the surviving rabbits being allowed to reproduce. The next generation of rabbits is likely to, on average, be better at running from the wolf. Similarly, in a search-based PCG implementation, a generation of strategy game units might be subject to selection by an evaluation function that grades them based on how complementary they are, and then mixed with each other (*recombination* or *crossover*) or copied with small random changes (*mutation*). The next generation of strategy game units is likely to, on average, be more complementary. It is important to note that this process works even when the initial generation consists of randomly generated individuals which are all very unfit for the purpose; some individuals will be less worthless than others, and a well-designed evaluation function will reflect these differences.

To make matters more concrete, let us describe a simple but fully usable evolutionary algorithm, the $\mu + \lambda$ *evolution strategy* (ES). The parameter μ represents the size of the part of the population that is kept between generations, the *elite*; the parameter λ represents the size of the part of the population that is generated through reproduction in each generation. For simplicity, imagine that $\mu = \lambda = 50$ while reading the following description.

1. Initialise the population of $\mu + \lambda$ individuals. The individuals could be randomly generated, or include some individuals that were hand-designed or the result of previous evolutionary runs.
2. Shuffle the population (permute it randomly). This phase is optional but helps in escaping loss-of-gradient situations.
3. Evaluate all individuals with the evaluation function, or some combination of several evaluation functions, so that each individual is assigned a single numeric value indicating its fitness.
4. Sort the population in order of ascending fitness.
5. Remove the λ worst individuals.
6. Replace the λ removed individuals with copies of the μ remaining individuals. The newly made copies are called the *offspring*. If $\mu = \lambda$, each individual in the elite is copied once; otherwise, it could be copied fewer or more times.
7. Mutate the λ offspring, i.e. perturb them randomly. The most suitable mutation operator depends on the representation and to some extent on the fitness landscape. If the representation is a vector of real numbers, an effective mutation operator is *Gaussian mutation*: add random numbers drawn from a Gaussian distribution with a small standard deviation to all numbers in the vector.
8. If the population contains an individual of sufficient quality, or the maximum number of generations is reached, stop. Otherwise, go to step 2 (i.e. start the next generation).

Despite the simplicity of this algorithm (it can be implemented in 10–20 lines of code), the $\mu + \lambda$ ES can be remarkably effective; even degenerate versions such as the $1 + 1$ ES can work well. However, the evolution strategy is just one of several types of evolutionary algorithms; another commonly used type is the genetic algorithm, which relies more on recombination and less on mutation, and which uses different selection mechanisms. There are also several types of stochastic search/optimisation algorithms that are not strictly speaking evolutionary algorithms but can be used for the same purpose, e.g. swarm intelligence algorithms such as particle swarm optimisation and ant colony optimisation. A good overview of evolutionary algorithms and some related approaches can be found in Eiben and Smith's book [8].

Some evolutionary algorithms are especially well suited to particular types of representation. For example, numerous variations on evolutionary algorithms have been developed especially for evolving runnable computer programs, often represented as expression trees [18]. If the artefacts are represented as vectors of real numbers of relatively short length (low dimensionality), a particularly effective algorithm is the Covariance Matrix Adaptation Evolution Strategy (CMA-ES), for which several open source implementations are available [9].

In many cases we want to use more than one evaluation function, as it is hard to capture all aspects of an artefact's quality in one number. In a standard single-objective evolutionary algorithm such as the evolution strategy, the evaluation functions could be combined as a weighted sum. However, this comes with its own set of problems, particularly that some functions tend to be optimised at the expense of others. Instead, one could use a *multiobjective* evolutionary algorithm, that optimises for several objectives at the same time and finds the set of *nondominated* individuals which have unique combinations of strengths. The most popular multiobjective evolutionary algorithm is perhaps the NSGA-II [7].

2.2.1 Other types of search algorithms

It could be argued that an evolutionary algorithm is "overkill" for some content generation problems. If your search space is very small and/or you have lots of time at hand to produce your content, you could try an exhaustive search algorithm that simply iterates through all possible configurations. In other cases, when it is easy to find good solutions and it is more important to maintain high diversity in the generated content, random search—simply sampling random points in the search space—could work well. Even when using exhaustive or random search the content needs to be represented in such a way that the space can be effectively searched/sampled and an evaluation function is necessary to tell the bad content from the good.

Another approach to content generation that can also be seen as search in content space is the solver-based approach, where e.g. Answer Set Programming is used to specify the logical conditions on game content. That approach will be discussed in Chapter 8.

2.3 Content representation

Content representation is a very important issue when evolving game content. The representation chosen plays an important role in the efficiency of the generation algorithm and the space of content the method will be able to cover. In evolutionary algorithms, the solutions in the generation space are usually encoded as *genotypes*, which are used for efficient searching and evaluation. Genotypes are later converted into *phenotypes*, the actual entities being evolved. In a game content generation scenario, the genotype might be the instructions for creating a game level, and the phenotype is the actual game level.

Examples of content representation in the game domain include the work done by Togelius et al. [26] who used an indirect representation to evolve maps for the real-time strategy game *StarCraft* [1]. In this experiment, the genotypes of maps were simply arrays of real numbers, whereas the phenotypes were complete StarCraft maps including passable/impassable areas, positions of bases and resources, etc. This experiment will be discussed in more detail in Section 2.5.

Fig. 2.1: A track evolved based on sequences of Bézier curves. Adapted from [24]

In another game genre, Cardamone et al. [4] evolved tracks for a car racing game. The tracks were represented as a set of control points the track has to cover and Bézier curves were employed to connect these points and ensure smoothness, a method inspired by the work done by Togelius et al. [24] on the same game genre. An example track evolved following this method is presented in Figure 2.1. This work will be discussed further in Section 2.6

As a concrete example of different representations, a level in *Super Mario Bros.* might be represented in any of the following ways.

1. Directly, as a level map, where each variable in the genotype corresponds to one "block" in the phenotype (e.g. bricks, question mark blocks, etc.).
2. More indirectly, as a list of the positions and properties of the different game entities such as enemies, platforms, gaps and hills (an example of this can be found in [19]).
3. Even more indirectly, as a repository of different reusable patterns (such as collections of coins or hills), and a list of how they are distributed (with various transforms such as rotation and scaling) across the level map (an example of this can be found in [23]).
4. Very indirectly, as a list of desirable properties such as number of gaps, enemies, coins, width of gaps, etc. (an example of this can be found in [20]).
5. Most indirectly, as a random number seed.

These representations yield very different search spaces. It's easy to think that the best representation would be the most direct one, which gives the evolutionary process most control over the phenotype. One should be aware, however, of the "curse of dimensionality" associated with representations that yield large search spaces: the larger the search space, the harder it is (in general) to find a certain solution. Another useful principle is that the representation should have a high *locality*, meaning that a small change to the genotype should on average result in a small change to the phenotype and a small change to the fitness value. In that sense, the last representation is unsuitable for search-based PCG because there is no locality, in which case all search methods perform as badly (or as well) as random search.

The choice of proper representation depends on the type of problem one is trying to solve. In the work done by Shaker et al. [20], the levels of *Infinite Mario Bros.* [17], a public clone of the popular game *Super Mario Bros.* [14], are represented according to option 4 as a vector of integers; each level is parametrized by four selected content features with the intention of finding the best combination of these features that can be used to generate content that optimises a specific player's experience. In a later study by the same authors [19], a more expressive representation is used following option 2, in which the structure of the levels of the same game is described in a design grammar that specifies the type, position, and properties of each item to be placed in the level map. Grammatical evolution is then applied to the design grammar in order to evolve new level designs [15]. A set of design elements, following option 3, was proposed in [23], also on the same game, where levels were described as a list of design elements placed in 2D maps; in this study a standard genetic algorithm was used to evolve content.

An issue closely related to the representation on the direct–indirect continuum is the expressive range of the chosen representation. The expressive range is relative to a particular measure of it: one could measure the expressivity of a platform game level generator in terms of how many different configurations of blocks it could produce, but it would make more sense to measure some quality that is more relevant to the experience of playing the game as a human. For example, the four-feature vector representation used to represent *Infinite Mario Bros.* levels allows control of the generation over only the four dimensions chosen, and consequently the search space is bounded by the range of these four features. On the other hand, a generator with a wider expressive range was built when representing the possible level designs in a design grammar which imposes fewer constraints on the structures evolved.

Chapter 9 further discusses the issue of representation in search-based PCG, and gives additional examples of representations tailored to particular content generation needs.

2.4 Evaluation functions

Candidate solutions, encoded in a represention, are evaluated by an evaluation function, which assigns a score (a fitness value or evaluation value) to each candidate. This is essential for the search process; if we do not have a good evaluation function, the evolutionary process will not work as intended and will not find good content. In general, the evaluation function should be designed to model some desirable quality of the artefact, e.g. its playability, regularity, entertainment value, etc. The design of an evaluation function depends to a great extent on the designer and what she thinks are the important aspects that should be optimised and how to formulate that.

For example, there are many studies on evolving game content that is "fun" [25, 24, 20, 4]. This term, however, is not well defined, and is hard to measure and formalise. This problem has been approached by many authors from different perspectives. In some studies, fun is considered a function of player behaviour and is measured accordingly. An example of such a method can be found in the work done by Togelius et al. [25] for evolving entertaining car racing tracks. In this study, indicators of player performance, such as the average speed achieved, were used as a measure of the suitability of each evolved track for individual players. In another study by Shaker et al. [20], fun is measured through self reports by directly asking the players about their experience. In other studies [22], a game is considered fun if the content presented follows predefined patterns that specify regions in the game and alternate between segments of varying challenge. In this case, challenge is considered the primary cause of a fun experience.

In search-based PCG, we can distinguish between three classes of evaluation functions: direct, simulation-based, and interactive.

2.4.1 Direct evaluation functions

Direct evaluation functions map features extracted from the generated content to a content quality value and, in that sense, they base their fitness calculations directly on the phenotype representation of the content. Direct evaluation functions are fast to compute and often relatively easy to implement, but it is sometimes hard to devise a direct evaluation function for some aspects of game content. Example features include the placement of bases and resources in real-time strategy games [26] or the size of the ruleset in strategy games [12]. The mapping between features and fitness might be contingent on a model of the playing style, preferences or affective state of players. An example of this form of fitness is the study done by Shaker et al. [20, 21] for personalising player experience using models of players to give a measure of content quality.

Within direct evaluation functions, two major types are *theory-driven* and *data-driven* functions. Theory-driven functions are guided by intuition and/or qualitative theories of player experience. Togelius et al. [24] used this method to evaluate the tracks in a car racing game. The evaluation function derived is based on several theoretical studies of fun in games [6, 11] combined with the authors' intuition of what makes an entertraining track. Data-driven functions, on the other hand, are based on quantitative measures of player experience that approximate the mapping between the content presented and players' affective or cognitive states collected via questionnaires or physiological measurements [21, 31].

2.4.2 Simulation-based evaluation functions

Simulation-based evaluation functions use AI agents that play through the content generated and estimate its quality. Statistics are usually calculated about the agents' behaviour and playing style and used to score game content. The type of the evaluation task determines the area of proficiency of the AI agent. If content is evaluated on the basis of playability, e.g. the existence of a path from the start to the end in a maze or a level in a 2D platform game, then AI agents should be designed that excel in reaching the end of the game. On the other hand, if content is optimised to maximise particular player experience, then an AI agent that imitates human behaviour is usually adopted. An example study that implements a human-like agent for assessing content quality is presented in [24] where neural-network-based controllers are trained to drive like human players in a car racing game and then used to evaluate the generated tracks. Each track generated is given a fitness value according to playing-behaviour statistics calculated while the AI controller is playing. Another example of a simulation-based evaluation function is measuring the average fighting time of bots in a first-person shooter game [5].

An important distinction within simulation-based evaluation functions is between *static* and *dynamic* functions. Static evaluation functions assume that the agent behaviour is maintained during gameplay. A dynamic evaluation function, on the other hand, uses an agent that adapts during gameplay. In such agents, the fitness value can be dependent on learnability: how well and/or fast the agent learns to play the content that is being evaluated.

2.4.3 Interactive evaluation functions

Interactive functions evaluate content based on interaction with a human, so they require a human "in the loop". Examples of this method can be found in the work done by Hastings et al. [10], who implemented this approach by evaluating the quality of the personalised weapons evolved implicitly based on how often and how long the player chooses to use these weapons. Cardamone et al. [4] also used this form of evaluation to score racing tracks according to the users' reported preferences. The first case is an example of an *implicit* collection of data while players' preferences were collected *explicitly* in the second. The problem with explicit data collection is that, if not well integrated, it requires the gameplay session to be interrupted. This method however provides a reliable and accurate estimator of player experience, as opposed to implicit data collection, which is usually noisy and based on assumptions. Hybrid approaches are sometimes employed to mitigate the drawbacks of these two methods by collecting information across multiple modalities such as combining player behaviour with eye gaze and/or skin conductance. Example studies that use this approach can be found in [13, 21, 31].

2.5 Example: *StarCraft* maps

In two recent papers, Togelius et al. presented a search-based approach to generating maps for the classic real-time strategy game (RTS) *StarCraft* [26, 27]. Despite being released in the previous millennium, this game is still widely played and was until recently the focus of large tournaments broadcast on national TV in countries such as South Korea. The focus of the game is on building bases, collecting resources, and waging war with armies of units built using these bases. The maps of the game play a crucial role, as they constrain what strategies are possible through their distribution of paths, obstacles, resources, etc. Given the competitive nature of the game, it is very important that the maps are fair. Therefore, evaluation functions were designed to measure the fairness of the maps as well as their affordances for interesting and diverse strategies.

Representation: The maps are represented as vectors of real numbers (of around 100 dimensions). In the genotype-to-phenotype process, some of these numbers are interpreted directly as the positions of resources or base starting locations. Other numbers are interpreted as starting positions and parameters for a turtle-graphics-like procedure that "draws" impassable regions (walls, rocks, etc.) on the initially empty map. The result of the transformation is a two-dimensional array where each cell corresponds to a block in the *StarCraft* map format; this can then be automatically converted into a valid *StarCraft* map.

Evaluation: Eight different evaluation functions were developed that address base placement, resource placement and paths between bases. These evaluation functions are based mostly on calculations of free space in different areas of the map and on the shortest paths between different points as calculated by the A* algorithm, and the functions are thus direct (though, if you see the path calculations as abstract simulations of unit behaviour in the game, the functions can be seen as simulation-based). There are functions for evaluating whether bases are sufficiently fair from each other, whether there is enough space to grow a base, and whether there is equal access to nearby resources. One particularly complicated function is the choke-point function, which returns a higher value if the shortest path between two bases has a choke point, a narrow area a tactically skilled player can use to defend against superior attacking forces by using level geometry.

Algorithm: Given the number of evaluation functions, it seemed very complicated to combine all of them into a single objective. SMS-EMOA, a state-of-the-art multiobjective evolutionary algorithm, was therefore used to evolve combinations of two or three objectives (some additional objectives were also converted to constraints). It was found that there are partial conflicts between several objectives, meaning that it is impossible to find a map that maximises all of them, but certain combinations of objectives yield interesting and reasonably fair maps.

2.6 Example: Racing tracks

Togelius et al. evolved racing tracks to fit particular players' playing styles in a simple two-dimensional racing game [24]. This particular game had already been used for a series of experiments investigating how evolutionary algorithms could best be used to create neural networks that could play the game well, when the authors decided to see whether the same technique could be applied to evolve the tracks the car was racing on. The reasoning was that creating challenging opponent drivers for commercial racing games is actually quite easy, especially if you are allowed to "cheat" by giving the computer-controlled cars superior performance (and who would stop you?)—on the other hand, creating an interesting racing track is not trivial at all.

Representation: The tracks are represented as vectors of real numbers, which are interpreted as control points for b-splines, i.e. sequences of Bézier curves.

Evaluation: The tracks are meant to be personalised for individual players. Therefore, the first stage in evolving a track for a given player is to model the playing style of that player. This is done by teaching a neural network (via another evolutionary process) to drive like that player. Then a candidate track is evaluated in a simulation-based manner by letting the neural network driver drive on that track in lieu of the human player and investigate its performance. This information is used by three different evaluation functions that measure whether the track has appropriate challenge and diversity for the player.

Algorithm: Given that there are three different evaluation functions, there remains the problem of combining them. The algorithm used, *cascading elitism*, is similar to $\mu + \lambda$ ES but has several stages of selection to ensure appropriate selection pressure on all objectives.

2.7 Example: Board game rules

Browne and Maire demonstrated that it is possible to automatically generate complete board games of such quality that they can be sold as commercial products [3]. The system described, *Ludi*, is restricted to simple board games similar to Go, Othello and Connect Four, but does a remarkable job of exploring this search space. This example will be discussed further in Chapter 6.

Representation: The board games, including board layouts and rules, were represented as strings (which can be interpreted as expression trees) in a special-purpose game description language. This is a relatively high-level language, describing entire games in just a few lines.

Evaluation: The games were evaluated by playing them with a version of the minimax game-tree search algorithm, with an evaluation function that had been automatically tuned for each game. A number of values were extracted from the performance of the algorithm on the game, e.g. how long it took to finish the game, how often the game ended in a draw, how many of the rules were used etc. These values were combined using a weighted sum based on empirical investigations of the properties of successful board games.

Algorithm: A relatively standard genetic algorithm was used.

2.8 Example: *Galactic Arms Race*

Galactic Arms Race (GAR) is a space shooter video game where the player traverses the space in a space ship, shoots enemies, collects items and upgrades their ship. The game was first released in 2010 as a free research game and a commercial version of the game was released in 2012. The game is interesting from a research perspective because it incorporates online automatic personalised content generation in a well-chosen playable context. The main innovation of the game is in personalising the weapons used by the player through evolution. As the game is played, new particle weapons are automatically generated based on player behaviour.

Representation: Particle system weapons are controlled by neural networks evolved by a method called NeuroEvolution of Augmenting Topologies (NEAT) [10]. NEAT evolves the networks through complexification, meaning that it starts with a population of simple, small networks, and increases the complexity of network topologies over generations. Each weapon in the game is represented as a single network that controls the motion (velocity) and appearance (colour) of the particles given the particle's current position in the space. The evolution starts with a set of simple weapons that shoot only in a straight line.

Evaluation: During the game, a fitness value is assigned to each weapon based on how much the particular weapon is used by the player; weapons used by the player more often are assigned higher fitness values, and thus have higher probability of being evolved. The newly evolved weapons are then spawned into space for the player to pick up.

Algorithm: The whole game thus represents a collective, distributed evolutionary algorithm. This process allows the generation of unique weapons for each player, increasingly personalised as they play the game.

2.9 Lab exercise: Evolve a dungeon

Roguelike games are a type of games that use PCG for level generation; in fact, the runtime generation and thereafter the infinite supply of levels is a key feature of this genre. As in the original game *Rogue* from 1980, a roguelike typically lets you control an agent in a labyrinthine dungeon, collecting treasures, fighting monsters and levelling up. A level in such a game thus consists of rooms of different sizes containing monsters and items and connected by corridors. There are a number of standard constructive algorithms for generating roguelike dungeons [16], such as:

- Create the rooms first and then connect them by corridors; or
- Use maze generation methods to create the corridors and then connect adjacent sections to create rooms.

The purpose of this exercise is to allow you to understand the search-based approach through implementing a search-based dungeon generator. Your generator should evolve playable dungeons for an imaginary roguelike. The phenotype of the dungeons should be 2D matrices (e.g. size 50×50) where each cell is one of the following: free space, wall, starting point, exit, monster, treasure. It is up to you whether to add other possible types of cell content, such as traps, teleporters, doors, keys, or different types of treasures and monsters. One of your tasks is to explore different content representations and quality measures in the context of dungeon generation. Possible content representations include [28]:

- A grid of cells that can contain one of the different items including: walls, items, monsters, free spaces and doors;
- A list of walls with their properties including their position, length and orientation;
- A list of different reusable patterns of walls and free space, and a list of how they are distributed across the grid;
- A list of desirable properties (number of rooms, doors, monsters, length of paths and branching factor); or
- A random number seed.

There are a number of advantages and disadvantages to each of these representations. In the first representation, for example, a grid of size 100×100 would need to be encoded as a vector of length 10,000, which is more than many search algorithms can effectively handle. The last option, on the other hand, explores one-dimensional space but it has no locality.

Content quality can be measured directly by counting the number of unreachable rooms or undesired properties such as a corridor connected to a corner in a room or a room connected to too many corridors.

2.10 Summary

In search-based PCG, evolutionary computation or other stochastic search/optimisation algorithms are used to create game content. The content creation can be seen as a search for the content that best satisfies an evaluation function in a content space. When designing a search-based PCG solution, the two main issues are the content representation and the evaluation function. The same space of content phenotypes can be represented in several different ways in genotype space; in general, we can talk about the continuum from direct representations (where genotypes are similar to phenotypes) to indirect representations (where genotypes are much smaller than phenotypes). Indirect representations yield less control and potentially sparser coverage of content space, but often cope better with the curse of dimensionality. There are three types of evaluation functions: direct, simulation-based, and interactive. Direct evaluation functions are fast, simulation-based evaluation functions require an AI to play through part of the game and interactive evaluation functions require a human in the loop. Search-based PCG is currently very popular in academia and there are multiple published studies; a few complete games have been released incorporating this approach to PCG.

References

1. Blizzard Entertainment, Mass Media: (1998). StarCraft, Blizzard Entertainment and Nintendo
2. Boden, M.A.: The creative mind: Myths and mechanisms. Psychology Press (2004)
3. Browne, C., Maire, F.: Evolutionary game design. IEEE Transactions on Computational Intelligence and AI in Games, **2**(1), 1–16 (2010)
4. Cardamone, L., Loiacono, D., Lanzi, P.L.: Interactive evolution for the procedural generation of tracks in a high-end racing game. In: Proceedings of the 13th Annual Conference on Genetic and Evolutionary Computation, pp. 395–402. ACM (2011)
5. Cardamone, L., Yannakakis, G.N., Togelius, J., Lanzi, P.: Evolving interesting maps for a first person shooter pp. 63–72 (2011)
6. Csikszentmihalyi, M.: Flow: The Psychology of Optimal Experience. Harper & Row (1991)
7. Deb, K., Pratap, A., Agarwal, S., Meyarivan, T.: A fast and elitist multiobjective genetic algorithm: NSGA-II. Evolutionary Computation, IEEE Transactions on **6**(2), 182–197 (2002)
8. Eiben, A.E., Smith, J.E.: Introduction to Evolutionary Computing. Springer (2003)
9. Hansen, N., Ostermeier, A.: Completely derandomized self-adaptation in evolution strategies. Evolutionary Computation **9**(2), 159–195 (2001)
10. Hastings, E.J., Guha, R., Stanley, K.: Evolving content in the Galactic Arms Race video game. In: Proceedings of the 5th International Conference on Computational Intelligence and Games, pp. 241–248. IEEE (2009)
11. Koster, R.: A Theory of Fun for Game Design. Paraglyph Press (2004)
12. Mahlmann, T., Togelius, J., Yannakakis, G.N.: Modelling and evaluation of complex scenarios with the Strategy Game Description Language. In: Computational Intelligence and Games (CIG), 2011 IEEE Conference on, pp. 174–181. IEEE (2011)
13. Martinez, H., Yannakakis, G.N.: Mining multimodal sequential patterns: A case study on affect detection. In: Proceedings of the 13th International Conference in Multimodal Interaction. ACM (2011)
14. Nintendo Creative Department: (1985). Super Mario Bros., Nintendo
15. O'Neill, M., Ryan, C.: Grammatical evolution. IEEE Transactions on Evolutionary Computation **5**(4), 349–358 (2001)
16. PCG Wiki: Procedural content generation wiki. URL http://pcg.wikidot.com/
17. Persson, M.: Infinite Mario Bros. URL http://www.mojang.com/notch/mario/
18. Poli, R., Langdon, W.B., McPhee, N.F.: A Field Guide to Genetic Programming (2008)
19. Shaker, N., Nicolau, M., Yannakakis, G.N., Togelius, J., O'Neill, M.: Evolving levels for Super Mario Bros. using grammatical evolution. In: Proceedings of the IEEE Conference on Computational Intelligence and Games (CIG), pp. 304–311. IEEE (2012)
20. Shaker, N., Togelius, J., Yannakakis, G.N.: Towards automatic personalized content generation for platform games. In: Proceedings of the AAAI Conference on Artificial Intelligence and Interactive Digital Entertainment (AIIDE). AAAI (2010)
21. Shaker, N., Yannakakis, G.N., Togelius, J., Nicolau, M., ONeill, M.: Fusing visual and behavioral cues for modeling user experience in games. IEEE Transactions on Systems Man, and Cybernetics **43**(6), 1519–1531 (2012)
22. Sorenson, N., Pasquier, P.: The evolution of fun: Automatic level design through challenge modeling. In: Proceedings of the First International Conference on Computational Creativity (ICCCX), pp. 258–267. ACM (2010)
23. Sorenson, N., Pasquier, P., DiPaola, S.: A generic approach to challenge modeling for the procedural creation of video game levels. IEEE Transactions on Computational Intelligence and AI in Games **3**(3), 229–244 (2011)
24. Togelius, J., De Nardi, R., Lucas, S.: Towards automatic personalised content creation for racing games. In: IEEE Symposium on Computational Intelligence and Games, 2007. CIG 2007, pp. 252–259. IEEE (2007)
25. Togelius, J., Nardi, R.D., Lucas, S.M.: Making racing fun through player modeling and track evolution. In: Proceedings of the SAB'06 Workshop on Adaptive Approaches for Optimizing Player Satisfaction in Computer and Physical Games (2006)

26. Togelius, J., Preuss, M., Beume, N., Wessing, S., Hagelbäck, J., Yannakakis, G.N.: Multiob-
    jective exploration of the StarCraft map space. In: Proceedings of the IEEE Conference on
    Computational Intelligence and Games (CIG), pp. 265–272 (2010)
27. Togelius, J., Preuss, M., Beume, N., Wessing, S., Hagelbäck, J., Yannakakis, G.N., Grappiolo,
    C.: Controllable procedural map generation via multiobjective evolution. Genetic Program-
    ming and Evolvable Machines **14**(2), 245–277 (2013)
28. Togelius, J., Yannakakis, G.N., Stanley, K.O., Browne, C.: Search-based procedural content
    generation. In: Proceedings of EvoApplications. Springer LNCS (2010)
29. Woodbury, R.F.: Searching for designs: Paradigm and practice. Building and Environment
    26(1), 61–73 (1991)
30. Wright, W.: The future of content. Talk at the 2005 Game Developers Conference.
    http://www.gdcvault.com/play/1019981/The-Future-of-Content (2005)
31. Yannakakis, G.N., Hallam, J.: Entertainment modeling in physical play through physiology
    beyond heart-rate. Affective Computing and Intelligent Interaction pp. 254–265 (2007)

Chapter 3
Constructive generation methods for dungeons and levels

Noor Shaker, Antonios Liapis, Julian Togelius, Ricardo Lopes, and Rafael Bidarra

Abstract This chapter addresses a specific type of game content, the dungeon, and a number of commonly used methods for generating such content. These methods are all "constructive", meaning that they run in fixed (usually short) time, and do not evaluate their output in order to re-generate it. Most of these methods are also relatively simple to implement. And while dungeons, or dungeon-like environments, occur in a very large number of games, these methods can often be made to work for other types of content as well. We finish the chapter by talking about some constructive generation methods for *Super Mario Bros.* levels.

3.1 Dungeons and levels

A dungeon, in the real world, is a cold, dark and dreadful place where prisoners are kept. A dungeon, in a computer game, is a labyrinthine environment where adventurers enter at one point, collect treasures, evade or slay monsters, rescue noble people, fall into traps and ultimately exit at another point. This conception of dungeons probably originated with the role-playing board game *Dungeons and Dragons*, and has been a key feature of almost every computer role-playing game (RPG), including genre-defining games such as the *Legend of Zelda* series and the *Final Fantasy* series, and recent megahits such as *The Elder Scrolls V: Skyrim*. Of particular note is the "roguelike" genre of games which, following the original *Rogue* from 1980, features procedural runtime dungeon generation; the *Diablo* series is a high-profile series of games in this tradition. Because of this close relationship with such successful games, and also due to the unique control challenges in their design, dungeons are a particularly active and attractive PCG subject.

For the purposes of this chapter, we define adventure and RPG dungeon levels as labyrinthic environments, consisting mostly of interrelated challenges, rewards and puzzles, tightly paced in time and space to offer highly structured gameplay progressions [9]. An aspect which sets dungeons apart from other types of levels

© Springer International Publishing Switzerland 2016

N. Shaker et al., *Procedural Content Generation in Games*, Computational Synthesis and Creative Systems, DOI 10.1007/978-3-319-42716-4_3

31

is a sophisticated notion of gameplay pacing and progression. Although dungeon levels are open for free player exploration (more than, say, platform levels), this exploration has a tight bond with the progression of challenges, rewards, and puzzles, as designed by the game's designer. And in contrast to platform levels or race tracks, dungeon levels encourage free exploration while keeping strict control over gameplay experience, progression and pacing (unlike open worlds, where the player is more independent). For example, players may freely choose their own dungeon path among different possible ones, but never encounter challenges that are impossible for their current skill level (since the space to backtrack is not as open as, for example, a sandbox city). Designing dungeons is thus a sophisticated exercise of emerging a complex game space from predetermined desired gameplay, rather than the other way around.

In most adventure games and RPGs, dungeons structurally consist of several rooms connected by hallways. While originally the term 'dungeon' refers to a labyrinth of prison cells, in games it may also refer to caves, caverns, or human-made structures. Beyond geometry and topology, dungeons include non-player characters (e.g. monsters to slay, princesses to save), decorations (typically fantasy-based) and objects (e.g. treasures to loot).

Procedural generation of dungeons refers to the generation of the topology, geometry and gameplay-related objects of this type of level. A typical dungeon generation method consists of three elements:

1. A representational model: an abstract, simplified representation of a dungeon, providing a simple overview of the final dungeon structure.
2. A method for constructing that representational model.
3. A method for creating the actual geometry of a dungeon from its representational model.

Above, we distinguished dungeons from platform levels. However, there are also clear similarities between these two types of game level. Canonical examples of platform game levels include those in *Super Mario Bros.* and *Sonic the Hedgehog*; a modern-day example of a game in this tradition that features procedural level generation is *Spelunky*, discussed in the first chapter. Like dungeons, platform game levels typically feature free space, walls, treasures or other collectables, enemies and traps. However, in the game mechanics of platformers, the player agent is typically constrained by gravity: the agent can move left or right and fall down, but can typically only jump a small distance upwards. As a result, the interplay of platforms and gaps is an essential element in the vocabulary of platform game levels.

In this chapter, we will study a variety of methods for procedurally creating dungeons and platform game levels. Although these methods may be very disparate, they have one feature in common: they are all constructive, producing only one output instance per run, in contrast with e.g. search-based methods. They also have in common that they are fast; some are even successful in creating levels at runtime. In general, these methods provide (rather) limited control over the output and its properties. The degree of control provided is nowadays a very important characteristic of any procedural method. By "control" we mean the set of options that a designer (or programmer) has in order to purposefully steer the level-generation process, as well as the amount of effort that steering takes. Control also determines whether editing those options and parameters causes sensible output changes, i.e. the intuitive responsiveness of a generator. Proper control assures that a generator creates consistent results (e.g. playable levels), while maintaining both the set of desired properties and variability.

We will discuss several families of procedural techniques. For simplicity, each of these techniques will be presented in the context of a single content type, either dungeons or platform game levels. The first family of algorithms to be discussed in this chapter is space partitioning. Two different examples of how dungeons can be generated by space partitioning are given; the core idea is to recursively divide the available space into pieces and then connect these pieces to form the dungeon. This is followed by a discussion of agent-based methods for generating dungeons, with the core idea that agents dig paths into a primeval mass of matter. The next family of algorithms to be introduced is cellular automata, which turn out to be a simple and fast means of generating structures such as cave-like dungeons. Generative grammars, yet another family of procedural methods, are discussed next, as they can naturally capture higher-level dungeon design aspects. We then turn our attention to several methods that were developed for generating platform levels, some of which are applicable to dungeons as well. The chapter ends with a discussion of the platform level generation methods implemented in the commercial game *Spelunky* and the open-source framework *Infinite Mario Bros.*, and its recent offshoot *InfiniTux*. The lab exercise will have you implement at least one method from the chapter using the *InfiniTux* API.

3.2 Space partitioning for dungeon generation

True to its name, a space-partitioning algorithm yields a *space partition*, i.e. a subdivision of a 2D or 3D space into disjoint subsets, so that any point in the space lies in exactly one of these subsets (also called *cells*). Space-partitioning algorithms often operate hierarchically: each cell in a space partition is further subdivided by applying the same algorithm recursively. This allows space partitions to be arranged in a so-called *space-partitioning tree*. Furthermore, such a tree data structure allows for fast geometric queries regarding any point within the space; this makes space partitioning trees particularly important for computer graphics, enabling, for example, efficient raycasting, frustum culling and collision detection.

The most popular method for space partitioning is *binary space partitioning* (BSP), which recursively divides a space into two subsets. Through binary space partitioning, the space can be represented as a binary tree, called a *BSP tree*.

Different variants of BSP choose different splitting hyperplanes based on specific rules. Such algorithms include quadtrees and octrees: a quadtree partitions a two-dimensional space into four quadrants, and an octree partitions a three-dimensional space into eight octants. We will be using quadtrees on two-dimensional images as

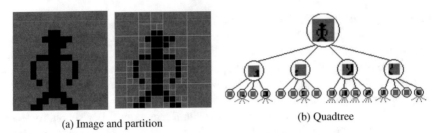

(a) Image and partition   (b) Quadtree

Fig. 3.1: Example quadtree partition of a binary image (0 shown as red, 1 as black). Large areas of a single colour, such as those on the right edge of the image, are not further partitioned. The image is 16 by 16 pixels, so the quadtree has a depth of 4. While a fully expanded quadtree (with leaf nodes containing information about a single pixel) would have 256 leaf nodes, the large areas of a single colour result in a quadtree with 94 leaf nodes. The first layers of the tree are shown in (b): the root node contains the entire image, with the four children ordered as: top left quadrant, top right quadrant, bottom left quadrant, bottom right quadrant (although other orderings are possible)

the simplest example. While a quadtree's quadrants can have any rectangular shape, they are usually equal-sized squares. A quadtree with a depth of n can represent any binary image of 2^n by 2^n pixels, although the total number of tree nodes and depth depends on the structure of the image. The root node represents the entire image, and its four children represent the top left, top right, bottom left, and bottom right quadrants of the image. If the pixels within any quadrant have different colours, that quadrant is subdivided; the process is applied recursively until each leaf quadrant (regardless of size) contains only pixels of the same colour (see Figure 3.1).

When space-partitioning algorithms are used in 2D or 3D graphics, their purpose is typically to represent existing elements such as polygons or pixels rather than to create new ones. However, the principle that space partitioning results in disjoint subsets with no overlapping areas is particularly suitable for creating rooms in a dungeon or, in general, distinct areas in a game level. Dungeon generation via BSP follows a *macro* approach, where the algorithm acts as an all-seeing dungeon architect rather than a "blind" digger as is often the case with the agent-based approaches presented in Section 3.3. The entire dungeon area is represented by the root node of the BSP tree and is partitioned recursively until a terminating condition is met (such as a minimum size for rooms). The BSP algorithm guarantees that no two rooms will be overlapping, and allows for a very structured appearance of the dungeon.

How closely the generative algorithms follow the principles of traditional partitioning algorithms affects the appearance of the dungeon created. For instance, a dungeon can be created from a quadtree by selecting quadrants at random and splitting them; once complete, each quadrant can be assigned a value of 0 (empty) or 1 (room), taking care that all rooms are connected. This creates very symmetric, 'square' dungeons such as those seen in Figure 3.2a. Furthermore, the principle that a leaf quadrant must consist of a uniform element (or of same-colour pixels,

(a)   (b)

Fig. 3.2: (a) A dungeon created using a quadtree, with each cell consisting entirely of empty space (black) or rooms (white). (b) A dungeon created using a quadtree, but with each quadrant containing a single room (placed stochastically) as well as empty space; corridors are added after the partitioning process is complete

in the case of images) can be relaxed for the purposes of dungeon generation; if each leaf quadrant contains a single room but can also have empty areas, this permits for rooms of different sizes, as long as their dimensions are smaller than the quadrant's bounds. These rooms can then be connected with each other, using random or rule-based processes, without taking the quadtree into account at all. Even with this added stochasticity, dungeons are still likely to be very neatly ordered (see Figure 3.2b).

We now describe an even more stochastic approach loosely based on BSP techniques. We consider an area for our dungeon, of width w and height h, stored in the root node of a BSP tree. Space can be partitioned along vertical or horizontal lines, and the resulting partition cells do not need to be of equal size. While generating the tree, in every iteration a leaf node is chosen at random and split along a randomly chosen vertical or horizontal line. A leaf node is not split any further if it is below a minimum size (we will consider a minimal width of $w/4$ and minimal height of $h/4$ for this example). In the end, each partition cell contains a single room; the corners of each room are chosen stochastically so that the room lies within the partition and has an acceptable size (i.e. is not too small). Once the tree is generated, corridors are generated by connecting children of the same parent with each other. Below is the high-level pseudocode of the generative algorithm, and Figures 3.3 and 3.4 show the process of generating a sample dungeon.

```
1: start with the entire dungeon area (root node of the BSP tree)
2: divide the area along a horizontal or vertical line
3: select one of the two new partition cells
4: if this cell is bigger than the minimal acceptable size:
5:    go to step 2 (using this cell as the area to be divided)
6: select the other partition cell, and go to step 4
7: for every partition cell:
8:    create a room within the cell by randomly
      choosing two points (top left and bottom right)
      within its boundaries
9: starting from the lowest layers, draw corridors to connect
   rooms corresponding to children of the same parent
```

in the BSP tree
10:repeat 9 until the children of the root node are connected

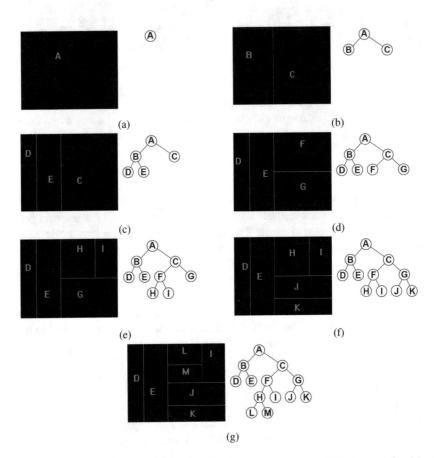

Fig. 3.3: Stochastically partitioning the dungeon area A, which is contained in the root node of the BSP tree. Initially the space is split into B and C via a vertical line (its x-coordinate is determined randomly). The smaller area B is split further with a vertical line into D and E; both D and E are too small to be split (in terms of width) so they remain leaf nodes. The larger area C is split along a horizontal line into F and G, and areas F and G (which have sufficient size to be split) are split along a vertical and a horizontal line respectively. At this point, the partition cells of G (J and K) are too small to be split further, and so is partition cell I of F. Cell H is still large enough to be split, and is split along a horizontal line into L and M. At this point all partitions are too small to be split further and dungeon partitioning is terminated with 7 leaf nodes on the BSP tree. Figure 3.4 demonstrates room and corridor placement for this dungeon

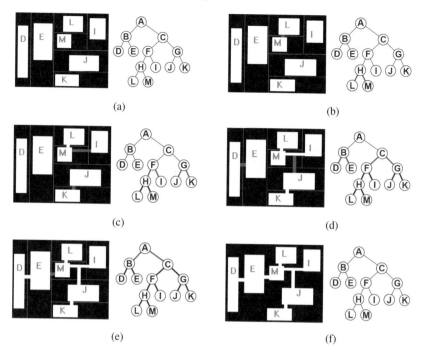

(a)   (b)

(c)   (d)

(e)   (f)

Fig. 3.4: Room and corridor placement in the partitioned dungeon of Figure 3.3. (a) For each leaf node in the BSP tree, a room is placed by randomly choosing coordinates for top left and bottom right corners, within the boundaries of the partition cell. (b) A corridor is added to connect the leaf nodes of the lowest layer of the tree (L and M); for all purposes, the algorithm will now consider rooms L and M as joined, grouping them together as their parent H. (c) Moving up the tree, H (the grouping of rooms L and M) is joined via a corridor with room I, and rooms J and K are joined via a corridor into their parent G. (d) Further up, rooms D and E of the same parent are joined together via a corridor, and the grouping of rooms L, M and I are joined with the grouping of rooms J and K. (e) Finally, the two subtrees of the root node are joined together and (f) the dungeon is fully connected

While binary space partitioning was primarily used here to create non-overlapping rooms, the hierarchy of the BSP tree can be used for other aspects of dungeon generation as well. The example of Figure 3.4 demonstrates how room connectivity can be determined by the BSP tree: using corridors to connect rooms corresponding to children of the same parent reduces the chances of overlapping or intersecting corridors. Moreover, non-leaf partition cells can be used to define groups of rooms following the same theme; for instance, a section of the dungeon may contain higher-level monsters, or monsters that are more vulnerable to magic. Coupled with corridor connectivity based on the BSP tree hierarchy, these groups of rooms may have a single entrance from the rest of the dungeon; this allows such a room to be decorated as a prison or as an area with dimmer light, favouring players who excel at stealthy gameplay. Some examples of themed dungeon partitions are shown in Figure 3.5.

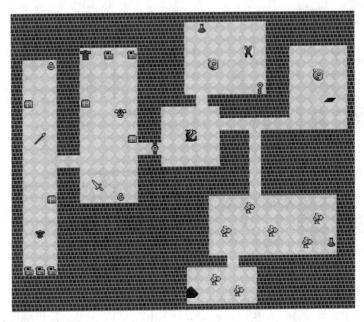

Fig. 3.5: The example dungeon from Figure 3.4, using the partitions to theme the room contents. Partition cells B and C are only connected by a single corridor; this allows the rooms of partition B to be locked away (green lock), requiring a key from cell C in order to be accessed (room L). Similarly, rooms of cell B contain only treasures and rewards, while rooms of partition C contain predominantly monsters. Moreover, the challenge rating of monsters in cell C is split between its child nodes: partition G contains weak goblins while cell F contains challenging monsters with magical powers. Further enhancements could increase the challenge of cell G by making it darker (placing fewer light sources), using different textures for the floor and walls of cell B, or changing the shape of rooms in cell C to circular

3.3 Agent-based dungeon growing

Agent-based approaches to dungeon generation usually amount to using a single agent to dig tunnels and create rooms in a sequence. Contrary to the space-partitioning approaches of Section 3.2, an agent-based approach such as this follows a *micro* approach and is more likely to create an organic and perhaps chaotic dungeon instead of the neatly organised dungeons of Section 3.2. The appearance of the dungeon largely depends on the behaviour of the agent: an agent with a high degree of stochasticity will result in very chaotic dungeons while an agent with some "look-ahead" may avoid intersecting corridors or rooms. The impact of the AI behaviour's parameters on the generated dungeons' appearance is difficult to guess without extensive trial and error; as such, agent-based approaches are much more unpredictable than space partitioning methods. Moreover, there is no guarantee that an agent-based approach will not create a dungeon with rooms overlapping each other, or a dungeon which spans only a corner of the dungeon area rather than its entirety. The following paragraphs will demonstrate two agent-based approaches for generating dungeons.

There is an infinite number of AI behaviours for digger agents when creating dungeons, and they can result in vastly different results. As an example, we will first consider a highly stochastic, 'blind' method. The agent is considered to start at some point of the dungeon, and a random direction is chosen (up, down, left or right). The agent starts digging in that direction, and every dungeon tile dug is replaced with a 'corridor' tile. After making the first 'dig', there is a 5% chance that the agent will change direction (choosing a new, random direction) and another 5% chance that the agent will place a room of random size (in this example, between three and seven tiles wide and long). For every tile that the agent moves in the same direction as the previous one, the chance of changing direction increases by 5%. For every tile that the agent moves without a room being added, the chance of adding a room increases by 5%. When the agent changes direction, the chance of changing direction again is reduced to 0%. When the agent adds a room, the chance of adding a room again is reduced to 0%. Figure 3.6 shows an example run of the algorithm, and its pseudocode is below.

```
1: initialize chance of changing direction Pc=5
2: initialize chance of adding room Pr=5
3: place the digger at a dungeon tile and randomize its direction
4: dig along that direction
5: roll a random number Nc between 0 and 100
6: if Nc below Pc:
7:   randomize the agent's direction
8:   set Pc=0
9: else:
10:  set Pc=Pc+5
11:roll a random number Nr between 0 and 100
12:if Nr below Pr:
13:  randomize room width and room length between 3 and 7
14:  place room around current agent position
14:  set Pr=0
15:else:
16:  set Pr=Pr+5
17:if the dungeon is not large enough:
18:  go to step 4
```

In order to avoid the lack of control of the previous stochastic approach, which can result in overlapping rooms and dead-end corridors, the agent can be a bit more

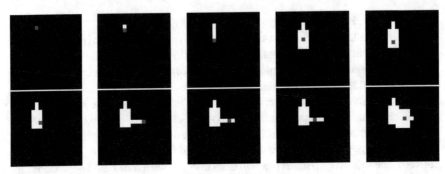

Fig. 3.6: A short run of the stochastic, "blind" digger. The digger starts at a random tile on the map (1st image), and starts digging downwards. After digging 5 tiles (3rd image), the chance of adding a room is 25%, and it is rolled, resulting in the 4th image. The agent continues moving downwards (4th image) with the chance of adding a room at 5% and the chance of changing direction at 30%: it is rolled, and the new direction is right (6th image). After moving another 5 tiles (7th image), the chance of adding a room is at 30% and the chance of changing direction is at 25%. A change of direction is rolled, and the agent starts moving left (8th image). After another tile is dug (9th image), the chance of adding a room is 40% and it is rolled, causing a new room to be added (10th image). Already, from this very short run, the agent has created a dead-end corridor and two overlapping rooms

informed about the overall appearance of the dungeon and look ahead to see whether the addition of a room would result in room–room or room–corridor intersections. Moreover, the change of direction does not need to be rolled in every step, to avoid winding pathways.

We will consider a less stochastic agent with look-ahead as a second example. As above, the agent starts at a random point in the dungeon. The agent checks whether adding a room in the current position will cause it to intersect existing rooms. If all possible rooms result in intersections, the agent picks a direction and a digging distance that will not result in the potential corridor intersecting with existing rooms or corridors. The algorithm stops if the agent stops at a location where no room and no corridor can be added without causing intersections. Figure 3.7 shows an example run of the algorithm, and below is its pseudocode.

```
1: place the digger at a dungeon tile
2: set helper variables Fr=0 and Fc=0
3: for all possible room sizes:
3:   if a potential room will not intersect existing rooms:
4:     place the room
5:     Fr=1
6:     break from for loop
7: for all possible corridors of any direction and length 3 to 7:
8:   if a potential corridor will not intersect existing rooms:
9:     place the corridor
10:     Fc=1
```

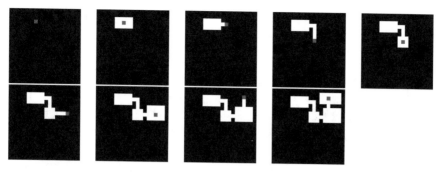

Fig. 3.7: A short run of the informed, "look ahead" digger. The digger starts at a random tile on the map (1st image), and places a room (2nd image) and a corridor (3rd image) since there can't be any overlaps in the empty dungeon. After placing the first corridor, there is no space for a room (provided rooms must be at least 3 by 3 tiles) which doesn't overlap with the previous room, so the digger makes another corridor going down (4th image). At this point, there is space for a small room which doesn't overlap (5th image) and the digger carries on placing corridors (6th image and 8th image) and rooms (7th image and 9th image) in succession. After the 9th image, the digger can't add a room or a corridor that doesn't intersect with existing rooms and corridors, so generation is halted despite a large part of the dungeon area being empty

```
11:    break from for loop
12:if Fr=1 or Fc=1:
13:  go to 2
```

The examples provided with the "blind" and "look-ahead" digger agents show naive, simple approaches; Figures 3.6 and 3.7 show to a large degree worst-case scenarios of the algorithm being run, with resulting dungeons either overlapping or being prematurely terminated. While simpler or more complex code additions to the provided digger behaviour can avert many of these problems, the fact still remains that it is difficult to anticipate such problems without running the agent's algorithm on extensive trials. This may be a desirable attribute, as the uncontrollability of the algorithm may result in organic, realistic caves (simulating human miners trying to tunnel their way towards a gold vein) and reduce the dungeon's predictability to a player, but it may also result in maps that are unplayable or unentertaining. More than most approaches presented in this chapter, the digger agent's parameters can have a very strong impact on the playability and entertainment value of the generated artefact and tweaking such parameters to best effect is not a straightforward or easy task.

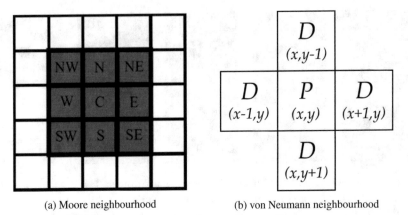

(a) Moore neighbourhood   (b) von Neumann neighbourhood

Fig. 3.8: Two types of neighbourhoods for cellular automata. Adapted from Wikipedia

3.4 Cellular automata

A cellular automaton (plural: cellular automata) is a discrete computational model. Cellular automata are widely studied in computer science, physics and even some branches of biology, as models of computation, growth, development, physical phenomena, etc. While cellular automata have been the subject of many publications, the basic concepts are actually very simple and can be explained in a few paragraphs and a picture or two.

A cellular automaton consists of an n-dimensional grid, a set of states and a set of transition rules. Most cellular automata are either one-dimensional (vectors) or two-dimensional (matrices). Each cell can be in one of several states; in the simplest case, cells can be *on* or *off*. The distribution of cell states at the beginning of an experiment (at time t_0) is the initial state of the cellular automaton. From then on, the automaton evolves in discrete steps based on the rules of that particular automaton. At each time t, each cell decides its new state based on the state of itself and all of the cells in its *neighbourhood* at time $t - 1$.

The neighbourhood defines which cells around a particular cell c affect c's future state. For one-dimensional cellular automata, the neighbourhood is defined by its size, i.e. how many cells to the left or right the neighbourhood stretches. For two-dimensional automata, the two most common types of neighbourhoods are *Moore neighbourhoods* and *von Neumann neighbourhoods*. Both neighbourhoods can have a size of any whole number, one or greater. A Moore neighbourhood is a square: a Moore neighbourhood of size 1 consists of the eight cells immediately surrounding c, including those surrounding it diagonally. A von Neumann neighbourhood is like a cross centred on c: a von Neumann neighbourhood of size 1 consists of the four cells surrounding c above, below, to the left and to the right (see Figure 3.8).

The number of possible configurations of the neighbourhood equals the number of states for a cell to the power of the number of cells in the neighbourhood. These numbers can quickly become huge, for example a two-state automaton with a Moore neighbourhood of size 2 has $2^{25} = 33,554,432$ configurations. For small neighbourhoods, it is common to define the transition rules as a table, where each possible configuration of the neighbourhood is associated with one future state, but for large neighbourhoods the transition rules are usually based on the proportion of cells that are in each state.

Cellular automata are very versatile, and several types have been shown to be Turing complete. It has even been argued that they could form the basis for a new way of understanding nature through bottom-up modelling [28]. However, in this chapter we will mostly concern ourselves with how they can be used for procedural content generation.

Johnson et al. [4] describe a system for generating infinite cave-like dungeons using cellular automata. The motivation was to create an infinite cave-crawling game, with environments stretching out endlessly and seamlessly in every direction. An additional design constraint is that the caves are supposed to look organic or eroded, rather than having straight edges and angles. No storage medium is large enough to store a truly endless cave, so the content must be generated at runtime, as players choose to explore new areas. The game does not scroll but instead presents the environment one screen at a time, which offers a time window of a few hundred milliseconds in which to create a new room every time the player exits a room.

This method uses the following four parameters to control the map generation process:

- A percentage of rock cells (inaccessible areas);
- The number of cellular automata generations;
- A neighbourhood threshold value that defines a rock (*T=5*);
- The number of neighbourhood cells.

Each room is a 50×50 grid, where each cell can be in one of two states: *empty* or *rock*. Initially, the grid is empty. The generation of a single room works as follows.

- The grid is "sprinkled" with rocks: for each cell, there is probability r (e.g. 0.5) that it is turned into rock. This results in a relatively uniform distribution of rock cells.
- A cellular automaton is applied to the grid for n (e.g. 2) steps. The single rule of this cellular automaton is that a cell turns into rock in the next time step if at least T (e.g. 5) of its neighbours are rock, otherwise it will turn into free space.
- For aesthetic reasons the rock cells that border on empty space are designated as "wall" cells, which are functionally rock cells but look different.

This simple procedure generates a surprisingly lifelike cave-room. Figure 3.9 shows a comparison between a random map (sprinkled with rocks) and the results of a few iterations of the cellular automaton.

But while this generates a single room, the game requires a number of connected rooms. A generated room might not have any openings in the confining rocks, and

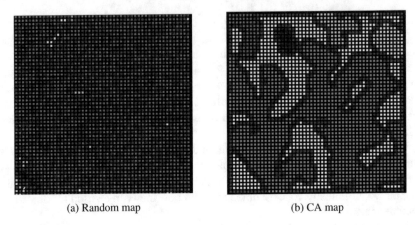

(a) Random map   (b) CA map

Fig. 3.9: Cave generation: Comparison between a CA and a randomly generated map ($r = 0.5$ in both maps); CA parameters: $n = 4$, $M = 1$, $T = 5$. Rock and wall cells are represented by white and red colour respectively. Coloured areas represent different tunnels (floor clusters). Adapted from [4]

there is no guarantee that any exits align with entrances to the adjacent rooms. Therefore, whenever a room is generated, its immediate neighbours are also generated. If there is no connection between the largest empty spaces in the two rooms, a tunnel is drilled between those areas at the point where they are least separated. Two more iterations of the cellular automaton are then run on all nine neighbouring rooms together, to smooth out any sharp edges. Figure 3.10 shows the result of this process, in the form of nine rooms that seamlessly connect. This generation process is extremely fast, and can generate all nine rooms in less than a millisecond on a modern computer.

We can conclude that the small number of parameters, and the fact that they are relatively intuitive, is an asset of cellular automata approaches like Johnson et al.'s. However, this is also one of the downsides of the method: for both designers and programmers, it is not easy to fully understand the impact that a single parameter has on the generation process, since each parameter affects multiple features of the generated maps. It is not possible to create a map that has specific requirements, such as a given number of rooms with a certain connectivity. Therefore, gameplay features are somewhat disjoint from these control parameters. Any link between this generation method and gameplay features would have to be created through a process of trial and error.

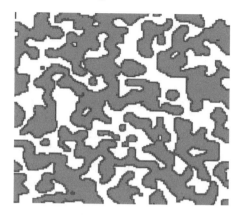

Fig. 3.10: Cave generation: a 3×3 base grid map generated with CA. Rock and wall cells are represented by white and red colour respectively. Grey areas represent floor. ($M = 2; T = 13; n = 4; r = 50\%$). Adapted from [4]

3.5 Grammar-based dungeon generation

Generative grammars were originally developed to formally describe structures in natural language. These structures—phrases, sentences, etc.—are modelled by a finite set of recursive rules that describe how larger-scale structures are built from smaller-scale ones, grounding out in individual words as the terminal symbols. They are *generative* because they describe linguistic structures in a way that also describes how to generate them: we can sample from a generative grammar to produce new sentences featuring the structures it describes. Similar techniques can be applied to other domains. For example, graph grammars [15] model the structure of graphs using a similar set of recursive rules, with individual graph nodes as the terminal symbols.

Back to our topic of dungeon generation, Adams [1] uses graph grammars to generate first-person shooter (FPS) levels. FPS levels may not obviously be the same as dungeons, but for our purposes his levels qualify as dungeons, because they share the same structure, a maze of interconnected rooms. He uses the rules of a graph grammar to generate a graph that describes a level's topology: nodes represent rooms, and an edge between two rooms means that they are adjacent. The method doesn't itself generate any further geometric details, such as room sizes. An advantage of this high-level, topological representation of a level is that graph generation can be controlled through parameters such as difficulty, fun, and global size. A search algorithm looks for levels that match input parameters by analyzing all results of a production rule at a given moment, and selecting the rule that best matches the specified targets.

One limit of Adams' work is the ad-hoc and hard-coded nature of its grammar rules, and especially the parameters. It is a sound approach for generating the topological description of a dungeon, but generalizing it to a broader set of games and goals would require creating new input parameters and rules each time. Regardless, Adams' results showcase the motivation and importance of controlling dungeon generation through gameplay.

Dormans' work [3] is more extensively covered in Chapter 5, so here we only briefly refer to his use of generative grammars to generate dungeon spaces for adventure games. Through a graph grammar, missions are first generated in the form of a directed graph, as a model of the sequential tasks that a player needs to perform. Subsequently, each mission is abstracted to a network of nodes and edges, which is then used by a shape grammar to generate a corresponding game space.

This was the first method to successfully introduce gameplay-based control, most notably with the concept of a mission grammar. Still, the method does not offer real control parameters, since control is actually exerted by the different rules in the graph and shape grammars, which are far from intuitive for most designers.

Inspired by the work of Dormans, van der Linden et al. [8] proposed the use of gameplay grammars to generate dungeon levels. Game designers express a-priori design constraints using a gameplay-oriented vocabulary, consisting of player actions to perform in-game, their sequencing and composition, inter-relationships and associated content. These designer-authored constraints directly result in a generative graph grammar, a so-called *gameplay grammar*, and multiple grammars can be expressed through different sets of constraints. A grammar generates graphs of player actions, which subsequently determine layouts for dungeon levels. For each generated graph, specific content is synthesized by following the graph's constraints. Several proposed algorithms map the graph into the required game space and a second procedural method generates geometry for the rooms and hallways, as required by the graph.

This approach aims at improving gameplay-based control on a generic basis, as it provides designers with the tools to effectively create, from scratch, grammar-based generators of graphs of player actions. The approach is generic, in the sense that such tools are not connected to any domain, and player actions and related design constraints can be created and manipulated across different games. However, integration of graphs of player actions in an actual game requires a specialized generator, able to transform such a graph into a specific dungeon level for that game. Van der Linden et al. demonstrated such a specialized generator for only one case study, yielding fully playable 3D dungeon levels for the game *Dwarf Quest* [27]. Figure 3.11 show (a) a gameplay graph and (b) a dungeon generated from this method.

As for gameplay-based control, this approach empowers designers to specify and control dungeon generation with a more natural design-oriented vocabulary. Designers can create their own player actions and use them as the vocabulary to control and author the dungeon generator. For this, they specify the desired gameplay which then constrains game-space creation. Furthermore, designers can express their own parameters (e.g. difficulty), which control rule rewriting in the gameplay grammar. Setting such gameplay-based parameters allows for even more fine-grained control over generated dungeons.

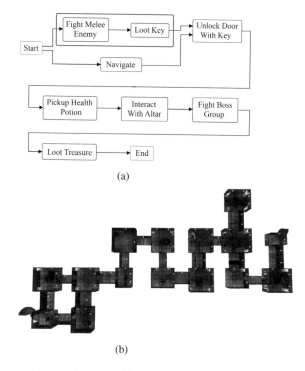

(a)

(b)

Fig. 3.11: (a) A gameplay graph created by van der Linden et al. [8] and (b) a corresponding dungeon layout generated for it

3.6 Advanced platform generation methods

In this section, we turn our attention to platform generation methods, by discussing two recent methods that were originally proposed for generating platform levels. Unlike the previous sections, there is no single category or family to characterize these methods. Interestingly, as we will point out, the central concepts of each of them could very well contribute to improve the generation of dungeons as well.

The first method, proposed by Smith et al. [23], is *rhythm-based platform generation*. It proposes level generation based on the notion of rhythm, linked to the timing and repetition of user actions. They first generate small pieces of a level, called rhythm groups, using a two-layered grammar-based approach. In the first layer, a set of player actions is created, after which this set of actions is converted into corresponding geometry. Many levels are created by connecting rhythm groups, and a set of implemented critics selects the best level.

Smith et al. propose a set of 'knobs' that a designer can manipulate to control the generation process, including (i) a general path through the level (i.e. start, end, and intermediate line segments), (ii) the kinds of rhythms to be generated, (iii) the types and frequencies of geometry components, and (iv) the way collectables (coins) are divided over the level (e.g. coins per group, probability for coins above gaps, etc.). There are also some parameters per created rhythm group, such as the frequency of jumps per rhythm group, and how often specific geometry (springs) should occur for a jump. Another set of parameters provides control over the rhythm length, density, beat type, and beat pattern.

The large number of parameters at different levels of abstraction provides many control options, and allows for the versatile generation of very disparate levels. Furthermore, they relate quite seamlessly to gameplay, especially in the platformer genre. However, this approach could nicely tie in with dungeon generation as well. As with Dormans, a two-layered grammar is used, where the first layer considers gameplay (in this case, player actions) and the second game space (geometry). The notion of *rhythm* as defined by Smith et al. is not directly applicable to dungeons, but the pacing or tempo of going through rooms and hallways could be of similar value in dungeon-based games. The decomposition of a level into rhythm groups also connects very well with the possible division of a dungeon into dungeon-groups with distinct gameplay features such as pacing.

Our second method, proposed by Mawhorter et al. [11] is called Occupancy-Regulated Extension (ORE), and it directly aims at procedurally generating 2D platform levels. ORE is a general geometry assembly algorithm that supports human-design-based level authoring at arbitrary scales. This approach relies on pre-authored chunks of level as a basis, and then assembles a level using these chunks from a library. A chunk is referred to as level geometry, such as a single ground element, a combination of ground elements and objects, interact-able objects, etc. This differs from the rhythm groups introduced by Smith et al. [23], because rhythm groups are separately generated by a PCG method whilst the ORE chunks are pieces of manually created content in a library. The algorithm takes the following steps: (i) a random potential player location (occupancy) is chosen to position a chunk; (ii) a chunk is selected from a list of context-based compatible chunks; (iii) the new chunk is integrated with the existing geometry. This process continues until there are no potential player locations left, after which post-processing takes care of placing objects such as power-ups.

This framework is meant for general 2D platform games, so specific game elements and mechanics need to be filled in, and chunks need to be designed and added to a library. Versatile levels can only be generated given that a minimally interesting chunk library is used.

Mawhorter et al. do not mention specific control parameters for their ORE algorithm, but a designer still has some control. Firstly, the chunks in the library and their probability of occurrence are implicit parameters, i.e. they actually determine the level geometry and versatility, and possible player actions need to be defined and incorporated in the design of chunks. And above all, their mixed-initiative approach provides the largest amount of control one can offer, even from a gameplay-based perspective. However, taken too far, this approach could come too close to manually constructing a level, decreasing the benefits of PCG. In summary, much control can be provided by this method, but the generation process may still be not very efficient, as a lot of manual work seems to still be required for specific levels to be generated.

This ORE method proposes a mixed-initiative approach, where a designer has the option to place content before the algorithm takes over and generates the rest of the level. This approach seems very interesting also for dungeon generation, where an algorithm that can fill in partially designed levels would be of great value. Imagine a designer placing special event rooms and then having an algorithm add the other parts of the level that are more generic in nature. This mixed-initiative approach would increase both level versatility, and control for designers, while still taking work off their hands. Additionally, it would fit the principles of dungeon design, where special rooms are connected via more generic hallways. Also, using a chunk library fits well in the context of dungeon-level generation (e.g. combining sets of template rooms, junctions and hallways). However, 3D dungeon levels would typically require a much larger and more complex chunk library than 2D platform levels, which share a lot of similar ground geometry.

3.7 Example applications to platform generation

3.7.1 Spelunky

Spelunky is a 2D platform indie game originally created by Derek Yu in 2008 [29]. The PC version of the game is available for free. An updated version of the game was later released in 2012 for the Xbox Live Arcade with better graphics and more content. An enhanced edition was also released on PC in 2013. The gameplay in *Spelunky* consists of traversing the 2D levels, collecting items, killing enemies and finding your way to the end. To win the game, the player needs to have good skills in managing different types of resources such as ropes, bumps and money. Losing the game at any level requires the game to be restarted from the beginning.

The game consists of four groups of maps of increasing level of difficulty. Each set of levels has a distinguished layout and introduces new challenges and new types of enemies. An example level from the second set is presented in Figure 3.12.

The standout feature of *Spelunky* is the procedural generation of game content. The use of PCG allows the generation of endless variations of content that are unique in every playthrough.

Each level in *Spelunky* is divided into a 4×4 grid of 16 rooms with two rooms marking the start and the end of the level (see Figure 3.13) and corridors connecting adjacent rooms. Not all the rooms are necessarily connected; in Figure 3.13 there are some isolated rooms such as the ones at the top left and bottom left corners. In order to reach these rooms, the player needs to use bombs, which are a limited resource, to destroy the walls.

The layout of each room is selected from a set of predefined templates. An example template for one of the rooms presented in Figure 3.13 can be seen in Fig-

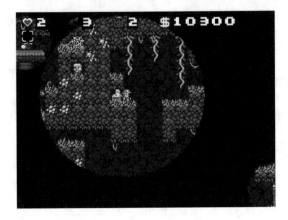

Fig. 3.12: Snapshot from *Spelunky*

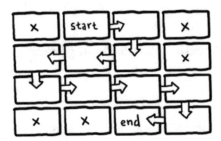

Fig. 3.13: Level generation in *Spelunky*. Adapted from [10]

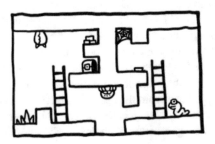

Fig. 3.14: Example room design in *Spelunky*. Adapted from [10]

ure 3.14. In each template, a number of chunks are marked in which randomisation can occur. Whenever a level is being generated, these chunks are replaced by different types of obstacle according to a set of randomised number generators [10]. Following this method, a new variation of the level can be generated with each run of the algorithm.

More specifically, each room in *Spelunky* consists of 80 tiles arranged in an 8×10 matrix [6]. An example room template can be:

```
0000000011
0060000L11
0000000L11
0000000L11
0000000L11
0000000011
0000000011
1111111111
```

Where 0 represents an empty cell, 1 stands for walls or bricks, L for ladders. The 6 in this example can be replaced by random obstacles permitting the generation of different variations. The obstacles, or traps, are usually of 5×3 blocks of tiles that overwrite the original tiles. Example traps included in the game can be spikes, webs or arrow traps, to name some.

While the basic layout of the level is partially random, with the presence of opportunities for variations, the placement of monsters and traps is 100% random. After generating the physical layout, the level map is scanned for potential places where monsters can be generated. These include, for example, a brick with empty tiles behind that offer enough space for generating a spider. There is another set of random numbers that controls the generation of monsters. These numbers control the type and the frequency of generation. For example, there is a 20% chance of creating a giant spider and once a spider is generated, this probability is set to 0 preventing the existence of more than one giant spider in a level.

In this sense, level generation in *Spelunky* can be seen as a composition of three main phases: in the first phase, the main layout of the level is generated by choosing the rooms from the templates available and defining the entrance and exit points. The second phase is obstacle generation, which can be thought of as an agent going through the level and placing obstacles in predefined spaces according to a set of heuristics. The final phase is the monster-generation phase, where another agent searches the level and places a monster when enough space is found and a set of conditions is satisfied.

3.7.2 Infinite Mario Bros.

Super Mario Bros. is a very popular 2D platform game developed by Nintendo and released in the mid 1980s [12]. A public domain clone of the game, named *Infinite Mario Bros.* (IMB) [14] was later published by Markus Persson. IMB features the art assets and general game mechanics of *Super Mario Bros.* but differs in level construction. IMB is playable on the web, where the Java source code is also available.[1] While implementing most features of *Super Mario Bros.*, the standout feature of IMB is the automatic generation of levels. Every time a new game is started, levels

---

[1] http://www.mojang.com/notch/mario/

are randomly generated by traversing the level map and adding features according to certain heuristics.

The internal representation of levels in IMB is a two-dimensional array of game elements. In "small" state, Mario is one block wide and one block high. Each position in the array can be filled with a brick block, a coin, an enemy or nothing. The levels are generated by placing the game elements in the two-dimensional level map.

Different approaches can be followed to generate the levels for this game [21, 19, 16, 24]. In the following we describe one possible approach.

The Probabilistic Multi-pass Generator (PMPG) was created by Ben Weber [21] as an entry for the level generation track of the Mario AI Championship [17]. The generator is agent-based and works by first creating the base level and then performing a number of passes through it. Level generation consists of six passes from left to right and adding one of the different types of game elements. Each pass is associated with a number of events (14 in total) that may occur according to predefined uniform probability distributions. These distributions are manually weighted and by tweaking these weights one can gain control over the frequency of different elements such as gaps, hills and enemies.

The six passes considered are:

1. An initial pass that changes the basic structure of the level by changing the height of the ground and starting or ending a gap;
2. the second pass adds the hills in the background;
3. the third pass adds the static enemies such as pipes and cannons based on the basic platform generated;
4. moving enemies such as koopas and goombas are added in the fourth pass;
5. the fifth pass adds the unconnected horizontal blocks, and finally,
6. the sixth pass places coins throughout the level.

Playability, or the existence of a path from the starting to the ending point, is guaranteed by imposing constraints on the items created and placed. For example, the width of generated gaps is limited by the maximum number of blocks that the player can jump over, and the height of pipes is limited to ensure that the player can pass through.

3.8 Lab session: Level generator for *InfiniTux* (and *Infinite Mario*)

InfiniTux, short for *Infinite Tux*, is a 2D platform game built by combining the underlying software used to generate the levels for *Infinite Mario Bros.* (IMB) with the art and sound assets of *Super Tux* [26]. The game was created to replace IMB, and is used in research [13, 22, 25, 7, 2] and the Mario AI Championship [20, 21, 5]. Since the level generator for *InfiniTux* is the same as the one used for IMB, the game features infinite variations of levels by the use of a random seed. The level of diffi-culty can also be tuned using different difficulty values, which control the number, frequency and types of the obstacles and monsters.

The purpose of this exercise is to use one or more of the methods presented in this chapter to implement your own generator that creates content for the game. The software you will be using is that used for the Level Generation Track of the Platformer AI Competition [18], a successor to the Mario AI Championship that is based on *InfiniTux*. The software provides an interface that eases interaction with the system and is a good starting point. You can either modify the original level generator, or use it as an inspiration. In order to help you to start with the software, we describe the main components of the interface provided and how it can be used.

As the software is developed for the Level Generation track of the competition, which invites participants to submit level generators that are fun for specific play-ers, the interface incorporates information about player behaviour that you could use while building your generator. This information is collected while the player is playing a test level and stored in a gameplay matrix that contains statistical features extracted from a gameplay session. The features include, for example, the number of jumps, the time spent running, the number of items collected and the number of enemies killed.

For your generator to work properly, your level should implement the *LevelIn-terface*, which specifies how the level is constructed and how different types of elements are scattered around the level:

```
public byte[][] getMap();
public SpriteTemplate[][] getSpriteTemplates()
```

The size of the level map is 320×15 and you should implement a method of your choice to fill in the map. Note that the basic structure of the level is saved in a different map than the one used to store the placement of enemies.

The level generator, which passes the gameplay matrix to your level and commu-nicates with the simulator, should implement the *LevelGenerator* interface:

```
public LevelInterface generateLevel(GamePlay playerMat);
```

There are quite a few examples reported in the literature that use this software for content creation; some of them are part of the Mario AI Championship and their implementation is open source and freely available at the competition website [17].

3.9 Summary

Constructive methods are commonly used for generating dungeons and levels in roguelike games and certain platformers, because such methods run in predictable, often short time. One family of such methods is based on binary space partitioning: recursively subdivide an area into ever smaller units, and then construct a dungeon by connecting these units in order. Another family of methods is based on agents that "dig out" a dungeon by traversing it in some way. While these methods originate in game development and might be seen as somewhat less principled, other methods for dungeon or level generation are applications of well-known computerscience techniques. Grammar-based methods, which are more extensively covered in Chapter 5, build dungeons by expanding from an axiom using production rules. Cellular automata are stochastic, iterative methods that can be used on their own or in combination with other methods to create smooth, organic-looking designs. Finally, several related methods work by going through a level in separate passes and adding content of different types according to simple rules with probabilities. Such methods have been used for the iconic roguelike platformer *Spelunky* and also for the Mario AI framework, but could easily be adapted to work for dungeons.

References

1. Adams, D.: Automatic generation of dungeons for computer games (2002). B.Sc. thesis, University of Sheffield, UK
2. Dahlskog, S., Togelius, J.: Patterns as objectives for level generation. In: Proceedings of the Second Workshop on Design Patterns in the 8th International Conference on the Foundations of Digital Games (2013)
3. Dormans, J.: Adventures in level design: generating missions and spaces for action adventure games. In: PCG'10: Proceedings of the 2010 Workshop on Procedural Content Generation in Games, pp. 1–8. ACM (2010)
4. Johnson, L., Yannakakis, G.N., Togelius, J.: Cellular automata for real-time generation of infinite cave levels. In: Proceedings of the 2010 Workshop on Procedural Content Generation in Games (2010)
5. Karakovskiy, S., Togelius, J.: The Mario AI benchmark and competitions. IEEE Transactions on Computational Intelligence and AI in Games **4**(1), 55–67 (2012)
6. Kazemi, D.: URL http://tinysubversions.com/2009/09/spelunkys-procedural-space/
7. Kerssemakers, M., Tuxen, J., Togelius, J., Yannakakis, G.: A procedural procedural level generator generator. In: IEEE Conference on Computational Intelligence and Games (CIG), pp. 335–341. IEEE (2012)
8. Van der Linden, R., Lopes, R., Bidarra, R.: Designing procedurally generated levels. In: Proceedings of the the 2nd AIIDE Workshop on Artificial Intelligence in the Game Design Process, pp. 41–47 (2013)
9. Van der Linden, R., Lopes, R., Bidarra, R.: Procedural generation of dungeons. IEEE Transactions on Computational Intelligence and AI in Games **6**(1), 78–89 (2014)
10. Make Games: URL http://makegames.tumblr.com/post/4061040007/the-full-spelunky-on-spelunky
11. Mawhorter, P., Mateas, M.: Procedural level generation using occupancy-regulated extension. In: IEEE Symposium on Computational Intelligence and Games (CIG), pp. 351–358 (2010)
12. Nintendo Creative Department: (1985). Super Mario Bros., Nintendo
13. Ortega, J., Shaker, N., Togelius, J., Yannakakis, G.N.: Imitating human playing styles in Super Mario Bros. Entertainment Computing pp. 93–104 (2012)
14. Persson, M.: Infinite Mario Bros. URL http://www.mojang.com/notch/mario/
15. Rozenberg, G. (ed.): Handbook of Graph Grammars and Computing by Graph Transformation: Volume I. Foundations. World Scientific (1997)
16. Shaker, N., Nicolau, M., Yannakakis, G.N., Togelius, J., O'Neill, M.: Evolving levels for Super Mario Bros. using grammatical evolution. In: IEEE Conference on Computational Intelligence and Games (CIG), pp. 304–311 (2012)

17. Shaker, N., Togelius, J., Karakovskiy, S., Yannakakis, G.: Mario AI Championship. URL http://marioai.org/
18. Shaker, N., Togelius, J., Yannakakis, G.: Platformer AI Competition. URL http://platformerai.com/
19. Shaker, N., Togelius, J., Yannakakis, G.N.: Towards automatic personalized content generation for platform games. In: Proceedings of the AAAI Conference on Artificial Intelligence and Interactive Digital Entertainment (AIIDE). AAAI (2010)
20. Shaker, N., Togelius, J., Yannakakis, G.N., Poovanna, L., Ethiraj, V.S., Johansson, S.J., Reynolds, R.G., Heether, L.K., Schumann, T., Gallagher, M.: The Turing test track of the 2012 Mario AI championship: Entries and evaluation. In: Proceedings of the IEEE Conference on Computational Intelligence and Games (CIG) (2013)
21. Shaker, N., Togelius, J., Yannakakis, G.N., Weber, B., Shimizu, T., Hashiyama, T., Sorenson, N., Pasquier, P., Mawhorter, P., Takahashi, G., Smith, G., Baumgarten, R.: The 2010 Mario AI championship: Level generation track. IEEE Transactions on Computational Intelligence and Games pp. 332–347 (2011)
22. Shaker, N., Yannakakis, G.N., Togelius, J., Nicolau, M., O'Neill, M.: Fusing visual and behavioral cues for modeling user experience in games. IEEE Transactions on Systems Man, and Cybernetics pp. 1519–1531 (2012)
23. Smith, G., Treanor, M., Whitehead, J., Mateas, M.: Rhythm-based level generation for 2D platformers. In: Proceedings of the 4th International Conference on Foundations of Digital Games, FDG 2009, pp. 175–182. ACM (2009)
24. Sorenson, N., Pasquier, P.: Towards a generic framework for automated video game level creation. In: Proceedings of the European Conference on Applications of Evolutionary Computation (EvoApplications), pp. 131–140. Springer LNCS (2010)
25. Sorenson, N., Pasquier, P., DiPaola, S.: A generic approach to challenge modeling for the procedural creation of video game levels. IEEE Transactions on Computational Intelligence and AI in Games (3), 229–244 (2011)
26. SuperTux Development Team: SuperTux. URL http://supertux.lethargik.org/
27. Wild Card: (2013). URL http://www.dwarfquestgame.com/
28. Wolfram, S.: Cellular Automata and Complexity: Collected Papers, vol. 1. Addison-Wesley Reading (1994)
29. Yu, D., Hull, A.: (2009). Spelunky

Chapter 4
Fractals, noise and agents with applications to landscapes

Noor Shaker, Julian Togelius, and Mark J. Nelson

Abstract Most games include some form of terrain or landscape (other than a flat floor) and this chapter is about how to effectively create the ground you (or the characters in your game) are standing on. It starts by describing several fast but effective stochastic methods for terrain generation, including the classic and widely used diamond-square and Perlin-noise methods. It then goes into agent-based methods for building more complex landscapes, and search-based methods for generating maps that include particular gameplay elements.

4.1 Terraforming and making noise

This chapter is about terrains (or landscapes—we will use the words interchangeably) and noise, two types of content which have more in common than might be expected. We will discuss three very different types of methods for generating such content, but first we will discuss where and why terrains and noise are used.

Terrains are ubiquitous. Almost any three-dimensional game will feature some ground to stand or drive on, and in most of them there will be some variety such as different types of vegetation, differences in elevation etc. What changes is how much you can interact directly with the terrains, and thus how they affect the game mechanics.

At one extreme of the spectrum are flight simulators. In many cases, the terrain has no game-mechanical consequences—you crash if your altitude is zero, but in most cases the minor variations in the terrain are not enough to affect your performance in the game. Instead, the role of the terrain is to provide a pretty backdrop and help the player to orientate. Key demands on the terrain are therefore that it is visually pleasing and believable, but also that it is huge: airplanes fly fast, are not hemmed in by walls, and can thus cover huge areas. From 30,000 feet one might not be able to see much detail and a low-resolution map might therefore be seen as a solution, but preferably it should be possible to swoop down close to the ground

© Springer International Publishing Switzerland 2016
N. Shaker et al., *Procedural Content Generation in Games*, Computational
Synthesis and Creative Systems, DOI 10.1007/978-3-319-42716-4_4

and see hills, houses, creeks and cars. Therefore, a map where the larger features were generated in advance but where details could be generated on demand would be useful. Also, from a high altitude it is easy to see the kind of regularities that result from essentially copying and pasting the same chunks of landscape, so reusing material is not trivial.

In open-world games such as *Skyrim* and the *Grand Theft Auto* series, terrains sometimes have mechanical and sometimes aesthetic roles. This poses additional demands on the design. When driving through a landscape in *Grand Theft Auto*, it needs to be believable and visually pleasing, but it also needs to support the stretch of road you are driving on. The mountains in *Skyrim* look pretty in the distance, but also function as boundaries of traversable space and to break line of sight. To make sure that these demands are satisfied, the generation algorithms need a high degree of controllability.

At the other end of the spectrum are those games where the terrain severely restricts and guides the player's possible course of actions. Here we find first-person shooters such as those in the *Halo* and *Call of Duty* series. In these cases, terrain generation has more in common with the level-generation problems we discussed in the previous chapter.

Like terrains, noise is a very common type of game content. Noise is useful whenever small variations need to be added to a surface (or something that can be seen as a surface). One example of noise is in skyboxes, where cloud cover can be implemented as a certain kind of white-coloured noise on a blue-coloured background. Other examples include dust that settles on the ground or walls, certain aspects of water (though water simulation is a complex topic in its own right), fire, plasma, skin and fur colouration etc. You can also see minor topological variations of the ground as noise, which brings us to the similarity between terrains and noise.

4.1.1 Heightmaps and intensity maps

Both noise and most aspects of terrains can fruitfully be represented as two-dimensional matrices of real numbers. The width and height of the matrix map to the x and y dimensions of a rectangular surface. In the case of noise, this is called an *intensity map*, and the values of cells correspond directly to the brightness of the associated pixels. In the case of terrains, the value of each cell corresponds to the height of the terrain (over some baseline) at that point. This is called a *heightmap*. If the resolution with which the terrain is rendered is greater than the resolution of the heightmap, intermediate points on the ground can simply be interpolated between points that do have specified height values. Thus, using this common representation, any technique used to generate noise could also be used to generate terrains, and vice versa—though they might not be equally suitable.

It should be noted that in the case of terrains, other representations are possible and occasionally suitable or even necessary. For example, one could represent the terrain in three dimensions, by dividing the space up into *voxels* (cubes) and computing the three-dimensional voxel grid. An example is the popular open-world game *Minecraft*, which uses unusually large voxels. Voxel grids allow structures that cannot be represented with heightmaps, such as caves and overhanging cliffs, but they require a much larger amount of storage.

4.2 Random terrain

Let's say we want to generate completely random terrain. We won't worry for the moment about the questions in the previous chapter, such as whether the terrain we generate would make a fair, balanced, and playable RTS map. All we want for now is random terrain, with no constraints except that it looks like terrain.

If we encode terrain as a heightmap, then it's represented by a two-dimensional array of values, which indicate the height at each point. Can generating random terrain be as simple as just calling a random-number generator to fill each cell of the array? Alas, no. While this technically works—a randomly initialized heightmap is indeed a heightmap that can be rendered as terrain—the result is not very useful. It doesn't look anything like random terrain, and isn't very useful as terrain, even if we're being generous. A random heightmap generated this way looks like random *spikes*, not random terrain: there are no flat portions, mountain ranges, hills, or other features typically identifiable on a landscape.

The key problem with just filling a heightmap with random values is that every random number is generated independently. In real terrain, heights at different points on the terrain are not independent of each other: the elevation at a specific point on the earth's surface is statistically related to the elevation at nearby points. If you pick a random point within 100 m of the peak of Mount Everest, it will almost certainly have a high elevation. If you pick a random point within 100 m of central Copenhagen, you are very unlikely to find a high elevation.

There are several alternative ways of generating random heightmaps to address this problem. These methods were originally invented, not for landscapes, but for textures in computer graphics, which had the same issue [3]. If we generate random graphical textures by randomly generating each pixel of the texture, this produces something that looks like television static, which isn't appropriate for textures that are going to represent the surfaces of "organic" patterns found in nature, such as the texture of rocks. We can think of landscape heightmaps as a kind of natural pattern, but a pattern that's interpreted as a 3D elevation rather than a 2D texture. So it's not a surprise that similar problems and solutions apply.

4.2.1 Interpolated random terrain

One way of avoiding unrealistically spiky landscapes is to require that the land-scapes we generate are smooth. That change does exclude some realistic kinds of landscapes, since discontinuities such as cliffs exist in real landscapes. But it's a change that will provide us with something much more landscape-like than the random heightmap method did.

How do we generate smooth landscapes? We might start by coming up with a formal definition of *smoothness* and then develop a method to optimise for that criterion. A simpler way is to make landscapes smooth by construction: fill in the values in such a way that the result is less spiky than the fully random generator. *Interpolated noise* is one such method, in which we generate fewer random values, and then interpolate between them.

With interpolated noise, instead of generating a random value at every point in the heightmap, we generate random values on a coarser *lattice*. The heights in between the generated lattice points are interpolated in a way that makes them smoothly connect the random heights. Put differently, we randomly generate elevations for peaks and valleys with a certain spacing, and then fill in the slopes between them.

That leaves one question: how do we do the interpolation, i.e. how do we connect the slopes between the peaks and valleys? There are a number of standard interpolation methods for doing so, which we'll discuss in turn.

4.2.1.1 Bilinear interpolation

A simple method of interpolating is to calculate a weighted average in first the horizontal, and then the vertical direction (or vice versa, which gives the same result). If we choose a lattice that's one-tenth as finely detailed as our heightmap's resolution, then $height[0,0]$ and $height[0,10]$ will be two of the randomly generated values. To fill in what should go in $height[0,1]$, then, we notice it's 10% of the way from $height[0,0]$ to $height[0,10]$. Therefore, we use the weighted average, $height[0,1] = 0.9 \times height[0,0] + 0.1 \times height[0,10]$. Once we've finished this interpolation in the x direction, then we do it in the y direction. This is called *bilinear interpolation*, because it does linear interpolation along two axes, and is both easy and efficient to implement.

While it's a simple procedure, coarse random generation on a lattice followed by bilinear interpolation does have drawbacks. The most obvious one is that mountain slopes become perfectly straight lines, and peaks and valleys are all perfectly sharp points. This is to be expected, since a geometric interpretation of the process just described is that we're randomly generating some peaks and valleys, and then filling in the mountain slopes by drawing straight lines connecting peaks and valleys to their neighbours. This produces a characteristically stylized terrain, like a child's drawing of mountains—perhaps what we want, but often not. For games in particular, we often don't want these sharp discontinuities at peaks and valleys, where collision detection can become wonky and characters can get stuck.

4.2.1.2 Bicubic interpolation

Rather than having sharp peaks and valleys connected by straight slopes, we can generate a different kind of stylized mountain profile. When a mountain rises from a valley, a common way it does so is in an S-curve shape. First, the slope starts rising slowly. It grows steeper as we move up the mountain; and finally it levels off at the top in a round peak. To produce this profile, we don't want to interpolate linearly: when we're 10% of the way between lattice points, we don't want to be 10% of the way up the slope's vertical distance yet.

Therefore we don't want to do a weighted average between the neighbouring lattice points according to their distance, but according to a nonlinear function of their distance. We introduce a slope function, $s(x)$, specifying how far up the slope (verically) we should be when we're x of the way between the lattice points, in the direction we're interpolating. In the bilinear interpolation case, $s(x) = x$. But now we want an $s(x)$ whose graph looks like an S-curve. There are many mathematical functions with that shape, but a common one used in computer graphics, because it's simple and fast to evaluate, is $s(x) = -2x^3 + 3x^2$. Now, when we are 10% of the way along, i.e. $x = 0.1$, $s(0.1) = 0.028$, so we should be only 2.8% up the slope's vertical height, still in the gradual portion at the bottom. We use this as the weight for the interpolation, and this time $height[0,1] = 0.972 \times height[0,0] + 0.028 \times height[0,10]$.

Since the $s(x)$ we chose is a cubic (third-power) function of x, and we again apply the interpolation in both directions along the 2D grid, this is called *bicubic interpolation*.

4.2.2 Gradient-based random terrain

In the examples so far, we've generated random values to put into the heightmap. Initially, we tried generating all the heightmap values directly, but that proved too noisy. Instead, we generated values for a coarse lattice, and interpolated the slopes in between the generated values. When done with bicubic interpolation, this produced a smooth slope.

An alternate idea is to generate the slopes directly, and infer height values from that, rather than generate height values and interpolate slopes. The random numbers we're going to generate will be interpreted as random *gradients*, i.e. the steepness and direction of the slopes. This kind of random initialization of an array is called *gradient noise*, rather than the *value noise* discussed in the previous section. It was first done by Ken Perlin in his work on the 1982 film *Tron*, so is sometimes called *Perlin noise*.

Generating gradients instead of height values has several advantages. Since we're interpolating gradients, i.e. rates of change in value, we have an extra level of smoothness: rather than smoothing the change in heights with an interpolation method, we smooth the *rate of change* in heights, so slopes grow shallower or steeper smoothly. Gradient noise also allows us to use lattice-based generation (which is computationally and memory efficient) while avoiding the rectangular grid effects produced by the interpolation-based methods. Since peaks and valleys are not directly generated on the lattice points, but rather emerge from the rises and falls of the slopes, they are arranged in a way that looks more organic.

As with interpolated value-based terrain, we generate numbers on a coarsely spaced lattice, and interpolate between the lattice points. However, we now generate a 2D vector, (d_x, d_y), at each lattice point, rather than a single value. This is the random gradient, and d_x and d_y can be thought of as the slope's steepness in the x and y directions. These gradient values can be positive or negative, for rising or falling slopes.

Now we need a way of recovering the height values from the gradients. First, we set the height to 0 at each lattice point. It might seem that this would produce noticeable grid artifacts, but unlike with value noise, it doesn't in practice. Since peaks and valleys rise and fall to different heights and with different slopes away from the $h = 0$ lattice points, the zero value is sometimes midway up a slope, sometimes near the bottom, and sometimes near the top, rather than in any visually regular position.

To find the height values at non-lattice points, we look at the four neighbouring lattice points. Consider first only the gradient to the top-left. What would the height value be at the current point if terrain rose or fell from $h = 0$ only according to that one of the four gradients? It would be simply that gradient's value multiplied by the distance we've traveled along it: the x-axis slope, d_x, times the distance we are to the right of the lattice point, added to the y-axis slope, d_y, times the distance we are down from the lattice point. In terms of vector arithmetic, this is the *dot product* between the gradient vector and a vector drawn from the lattice point to our current point.

Repeat this what-if process for each of the four surrounding lattice points. Now we have four height values, each indicating the height of the terrain if only one of the four neighbouring lattice points had influence on its height. Now to combine them, we simply interpolate these values, as we did with the value-noise terrain. We have four surrounding lattice points that now have four height values, and we have already covered, in the previous section, how to interpolate height values, using bilinear or bicubic interpolation.

4.3 Fractal terrain

While gradient noise looks more organic, there is still a rather unnatural aspect to it when treated as terrain: terrain now undulates at a constant frequency, which is the frequency chosen for the lattice point spacing. Real terrain has variation at multiple scales. At the largest scale (i.e. lowest frequency), plains rise into mountain ranges. But at smaller scales, mountain ranges have peaks and valleys, and valleys have smaller hills and ravines. In fact, as you zoom in to many natural phenomena, you see the same kind of variation that was seen at the larger scale, but reproduced at

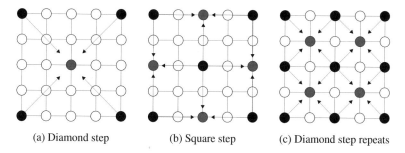

| (a) Diamond step | (b) Square step | (c) Diamond step repeats |

Fig. 4.1: The diamond-square algorithm. (Illustration credit: Amy Hoover)

a new, smaller scale [10]. This self-similarity is the basis of *fractals*, and generated terrain with this property is called *fractal terrain*.

Fractal terrain can be produced through a number of methods, some of them based directly on fractal mathematics, and others producing a similar effect via simpler means.

A very easy way to produce fractal terrain is to take the single-scale random terrain methods from the previous section and simply run them several times, at multiple scales. We first generate random terrain with very large-scale features, then with smaller-scale features, then even smaller, and add all the scales together. The larger-scale features are added in at a larger magnitude than the smaller ones: mountains rise from plains a larger distance than boulders rise from mountain slopes. A classic way of producing multi-scale terrain in this way is to scale the generated noise layers by the inverse of their frequency, which is called $1/f$ noise. If we have a single-scale noise-generation function, like those in the previous section, we can give it a parameter specifying the frequency; let's call this function $noise(f)$. Then starting from a base for our lowest-frequency (largest-scale) features, f, we can define $1/f$ noise as

$$noise(f) + \frac{1}{2}noise(2f) + \frac{1}{4}noise(4f) + \dots$$

There are many other methods for fractal terrain generation, most of which are beyond the scope of this book, as there exist other textbooks covering the subject in detail [3]. Musgrave et al. [11] group them into five categories of technical approaches, all of which can be seen as implementation methods for the general concept of fractional Brownian motion (fBm). In fBm, we can conceptually think of a terrain as being generated by starting from a point and then taking a random walk following specific statistical properties. Since actually taking millions of such random walks is too computationally expensive, a similar end result is approximated using a variety of techniques. One that is commonly used in games, because it is relatively simple to implement and computationally efficient, is the *diamond-square* algorithm, illustrated in Figure 4.1.

In the diamond-square algorithm, we start by setting the four corners of the heightmap to seed values (possibly random). The algorithm then proceeds as follows. First, find the point in the center of the square defined by these four corners, and set it to the average of the four corners' values plus a random value. This is the "diamond" step. Then find the four midpoints of the square's sides and set each of them to the average of three values: the two neighbouring corners and the middle of the square (which we just set in the last step)—again, plus a random value. This is the "square" step. The magnitude of the random values we use is called the *roughness*, because larger values produce rougher terrain (likewise, smaller values produce smoother terrain). Completing these two steps has subdivided the original square into four squares. We then reduce the roughness value and repeat the two steps, to fill in these smaller squares. Typically the process repeats until a specified maximum number of iterations have been reached. The end result is an approximation of the terrain produced by fBm.

4.4 Agent-based landscape creation

In Chapter 1, we discussed the desired properties of a PCG algorithm. The previously discussed methods satisfy most of these properties, however they suffer from uncontrollability. The results delivered by these methods are fairly random and they offer very limited interaction with designers, who can only provide inputs on the global level through modifying a set of unintuitive parameters [14]. Several variations of these methods have been introduced that grant more control over the output [7, 1, 13, 16].

The main advantage of software-agent approaches to terrain generation over fractal-based methods is that they offer a greater degree of control while maintaining the other desirable properties of PCG methods. Similarly to the agent-based approaches used in dungeon generation (Section 3.3), agent-based approaches for landscape creation grow landscapes through the action of one or more software agents. An example is the agent-based procedural city generation demonstrated by Lechner et al. [9]. In this work, cities are divided into areas (such as squares, industrial, commercial, residential, etc.) and agents construct the road networks. Different types of agents do different jobs, such as *extenders*, which search for unconnected areas in the city, and *connectors*, which add highways and direct connections between roads with long travel times. Later versions of this system introduced additional types of agents, for tasks such as constructing main roads and small streets [8].

But since this chapter is about terrain generation, we'll look now at work on agent-based terrain generation by Doran and Parberry [2], which focuses primarily on the issue of controllability, especially on providing more control to a designer than the dominant fractal-based terrain generation methods do. Because of the lack of input and interaction with designers, fractal-based methods are usually evaluated in term of efficiency rather than the aesthetic features of the terrains generated [2]. Agent-based approaches, on the other hand, offer the possibility of defining more fine-grained measures of the *goodness* of the terrains according to the behaviour of the agents. By controlling how and how much the agent changes the environment, one can vary the quality of the generated terrains.

4.4.1 Doran and Parberry's terrain generation

Doran and Parberry's terrain-generation approach starts with five different types of agents that work concurrently in an environment to simulate natural phenomena. The agents are allowed to sense the environment and change it at will. Designers are provided with a number of ways to influence terrain generation: controlling the number of agents of each type is one way to gain control, another is by limiting the agent lifetime using a predefined number of actions that the agent can perform. After the number of steps is consumed, the agent becomes inactive.

The agents can modify the environment by performing three main tasks:

- Coastline: in this phase, the outline or shape of the terrain is generated using multiple agents.
- Landform: the detailed features of the land are defined in this phase employing more agents than were used in the previous phase. The agents work simultaneously on the environment to set the details of the mountains, create beaches and shape the lowlands.
- Erosion: this is the last phase of the generation and it constitutes the creation of rivers through eroding the previously generated terrain. The number of river to create is determined by the number of agents defined in this phase.

According to these phases, several types of agents can be identified to achieve the several tasks defined in each phase. The authors focused their work on five different types:

1. Coastline agents: these agents work in the coastline phase before any other agents, to draw the outline of the landscape. The map is initially placed under sea level and the agents work by raising points above sea level. The process starts with a single agent working on the entire map. Depending on the size of the map, this agent multiplies by creating many other coastline agents which subdivide themselves in turn until each agent is assigned a small part of the map. The process undertaken by each agent to generate the coastline can be described as follows:

   - Each agent is assigned a single seed point at the edge of the map, a direction to follow and a number of tokens to consume.
   - The agent checks its surroundings and if it is already land (this might happen since all the agents are working simultaneously on the map) the agent starts searching in the assigned direction for another appropriate starting point.
   - Once the starting point is located, the agent starts working on the environment by changing the height of the points. This is done by

    a. generating two points at random in different directions: one works as an attractor and the other as a repulser.

    b. identifying the set of points for elevation above the sea level.

    c. scoring the points according to their distance from the attractor and the repulser points. The ones closer to the attractor are scored higher.

    d. the point with the highest score is then elevated above sea level and it becomes part of the coastline.

    e. the agent then continues by moving to another point in the map.

This method allows multiple agents to work concurrently on the map while preserving localization since each agent moves in its surroundings and has a predefined number of tokens to consume. The number of tokens given to each agent and the number of agents working on the map are directly related. The smaller the number of tokens, the larger the number of agents since more agents will be required to cover the whole map. These parameters also affect the level of detail of the coastline. A map generated with a small number of tokens will feature more fine details than one with a large number, since in the first case more agents will be created, each influencing a small region.

2. Smoothing agents: after the shape of the landscape has been defined by the coastline agents, smoothing agents operate on the map to eliminate rapid elevation changes. This is done by creating a number of agents each assigned a single parameter specifying the number of times that agent has to revisit its starting point. The more visits, the smoother the area around this point.

    The agents are scattered around the map, they move randomly and while wandering they change the heights of arbitrary points according to the heights of their neighbours. For each point chosen, a new height value is assigned taking the weighted averages of the heights of its four orthogonal surrounding points and the four points beyond these.

3. Beach agents: after the smoothing phase, the landscape is ready for the creation of sandy beaches. This is the work assigned to beach agents. These agents traverse the shoreline in random directions creating sandy areas close to water. Beach generation is controlled by adjusting the agents' parameters. These include the depth of the area the agent is allowed to flatten, the total number of steps the agents can move, the altitude under which the agents are permitted to work and the range of height values they can assign to the points they affect.

    The agents are initially placed in a coastline area where they work on adjusting the height of their surrounding points by lowering them as long as their height is below the predefined altitude. This prevents the elevation of mountain areas located close to the sea. The new values assigned to the points are randomly chosen from the designer-specified range. This allows the creation of flat beaches if the range is narrow and more bumpy beaches when the range is high.

4. Mountain agents: The coastline agents elevate areas of the map above sea level. These areas are then smoothed by the smoothing agents and beaches along the shoreline are then flattened via the beach agents. Regions above a certain thresh-

old are kept untouched by the beach agents, and these are then modified by mountain agents.

The agents are placed at random positions in the maps and are allowed to move in random directions. While moving, if a V-shaped wedge of points is encountered, the wedge is elevated, creating a ridge. Frequently, the agents might decide to turn randomly within 45 degrees of their initial course, resulting in zigzag paths. Mountain agents also periodically produce foothills perpendicular to their movement direction.

The shape of the mountains can be controlled by designers via specifying the range of the rate at which slopes can be dropped, the maximum mountain altitude and the width and slope of the mountain. Designers can also determine the number of agents, the number of steps each one can perform, the length of foothills and their frequency.

After mountain generation, a smoothing step is followed to blend nearby points. This step is further followed by an addition of noise to regain some of the details lost while smoothing.

5. Hill agents: these agents work in a similar way to the mountain agents but they have three distinctive characteristics: they work on a lower altitude, they are assigned smaller ranges, and they are not allowed to generate foothills.

6. River agents: in the final phase of terrain generation, river agents walk through the environment digging rivers near mountains and the ocean. To resemble natural rivers, a river agent works in the following steps:

   a. initiate two random points, one on the coastline and another on the mountain ridge line.
   b. starting at the coastline, the agent moves uphill towards a mountain, guided by the gradient. This determines the general path of the river.
   c. as the agent reaches the mountain, it starts moving downwards while digging the river. This is done by lowering a wedge of terrain, following a similar method to the one implemented by mountain agents.
   d. the agent increases the width of the wedge as it moves back towards the ocean.

Designers specify the initial width of the river, the frequency of widening and the downhill slope. Designers also determine the shortest length possible for a river. A river agent might make several attempts to place its starting and ending positions before it satisfies the shortest-length threshold. If this condition is not met after several attempts, the river will not be created.

The method followed for defining the agents and their set of parameters allows the generation of endless variations of terrains through the use of different random seed numbers. The technique can be used to generate landscapes on the fly, or it can be employed by designers who can investigate different setups and tweak the system's parameters as desired.

4.5 Search-based landscape generation

We have seen two families of methods for terrain and noise generation, which both
have several benefits. However, at least in the form presented here, these methods
suffer from a certain lack of controllability. It is not easy to specify constraints
or desirable properties, such that there must be an area with no more than a certain maximum variation in altitude, or that two points on a terrain should be easily
reachable from each other. This form of controllability is one of the strengths of
search-based methods. Unsurprisingly, there have been several attempts to apply
search-based methods to terrain generation.

4.5.1 Genetic terrain programming

Frade et al. developed a concept called genetic terrain programming (GTP). This is a
search-based method with an indirect encoding, where the phenotype representation
is a heightmap but the genotype representation is an *expression tree* evolved with
genetic programming [5, 6, 4].

Genetic programming is a method for creating runnable programs using evolutionary computation [12]. The standard program representation in genetic programming is an expression tree, which in the simplest case is nothing more than an alge-
braic expression in prefix form such as $(+3(*52))$ (written in infix form as $3+5*2$).
This can be visualized as a tree with the $+$ sign as the root node, and the 3 and $*$
in separate branches from the root. The plus and multiplier are arithmetical functions, and the constants are called *terminals*. In genetic programming, a number of
additional functions are commonly employed, including if-then-else, trigonometric
functions, max, min etc. Additional types of terminals might include external inputs
to the program, random-number generators etc. The evolutionary search proceeds
through adding and exchanging functions and terminals, and by recombining parts
of different trees.

In GTP, the function set typically includes arithmetical and trigonometric functions, as well as functions for exponentiation and logarithms. The terminal set includes x and y location, standard noise functions (such as Perlin noise) and functions
that are dependent on the distance from the centre of the map.

The core idea of GTP is that in the genotype-to-phenotype mapping, the algorithm iterates over cells in the (initially empty) heightmap and queries the evolved
terrain program with the x and y parameters of each cell as input to the program. This
is therefore a highly indirect and compact representation of the map. The representation also allows for infinite scalability (or zooming), as increasing the resolution
or expanding the map simply means querying the program using new coordinates
as inputs.

Several different evaluation functions were tried. In initial experiments, interactive evaluation was used: users selected which of several presented maps should be
used for generating the next generation. Later experiments explored various direct

evaluation functions. One of these functions, accessibility, was motivated by game design: the objective was to maximise the area which is smooth enough to support vehicle movement. To avoid completely flat surfaces from being evolved, the accessibility metric had to be counterbalanced by other metrics, such as the sum of the edge length of all obstacles in the terrain. See Figure 4.2 for some examples of landscapes evolved with GTP.

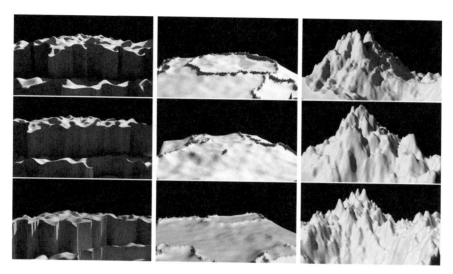

Fig. 4.2: Landscapes generated by genetic terrain programming. From left to right: cliffs, corals and mountains. Adapted from [4]

4.5.2 Simple RTS map generation

Another search-based landscape generation method was described by Togelius et al. [15], to produce a map with smoothly varying height for a real-time strategy game. The phenotype in this problem consists of a heightmap and the locations of resources and base starting locations.

The representation is rather direct. Base and resource locations are represented directly as polar coordinates (ϕ and θ coordinates for each location). The heightmap is initially flat, and then a number of hills is added. These hills are modelled as simple Gaussian distributions, and encoded in the phenotype with their x and y positions, their heights z, and their standard distributions σ_x and σ_y (i.e. their widths). Ten mountains were used in each run.

Three different evaluation functions were defined. Two of them relate to the placements of bases and resources to create a fair game, whereas the third is the topological asymmetry of the map. This is because the simplest way of satisfying

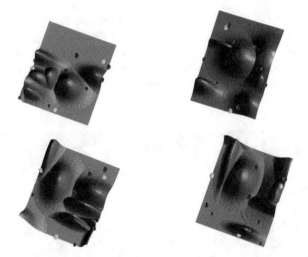

Fig. 4.3: Four maps generated using search-based methods with the heightmap determined by hills represented by Gaussians. The coloured dots represent locations of resources and bases in an RTS game. Adapted from [15]

the first two evaluation functions is to create a completely symmetric map, but this would be visually uninteresting for players. Given that the three fitness functions are in partial conflict, a multiobjective evolutionary algorithm was used to optimise all three evaluation functions simultaneously. Figure 4.3 show three different terrains that resulted from the same evolutionary run.

4.6 Lab session: Generate a terrain with the diamond-square algorithm

Implement the diamond-square method to generate terrain heightmaps. Have your function take three parameters: *seed*, which specifies the initial values at the corners; *iterations*, which specifies the number of diamond-square iterations to perform; and *roughness*, which specifies the magnitude of the random components added in the diamond and square steps.

Figure 4.4 presents three example heightmaps generated using different parameters.

Fig. 4.4: Three heightmaps generated using the diamond-square method. The parameters used are: iterations = 9 for all maps, seed = 12, 128, 128, and roughness = 256, 256, 128 for the first, second and third map, respectively

4.7 Summary

Terrain/landscape generation is a very important task for many games, and there are a number of methods that are commonly used for this. The most basic representation is the heightmap, where the number in each cell represents the height of the ground at the corresponding location. Maps can be generated very simply by randomizing these numbers, though this leads to unnatural and ugly maps. Interpolating between these numbers helps a lot. Many different interpolation techniques exist, and there is a tradeoff between the quality of the results and the computation time needed. Instead of generating the height values, another family of methods generates the gradients of slopes and then computes heights from those slopes. Fractal methods, including various types of noise, generate heights at several different scales or resolutions, leading to more natural-looking terrain. The diamond-square algorithm is a commonly used fractal method. For more complex environments, agent-based methods can be used to construct terrains that have multiple types of features. If there are constraints involved, for example having to do with traversability or other gameplay considerations, search-based methods might be useful as well.

References

1. Belhadj, F.: Terrain modeling: A constrained fractal model. In: Proceedings of the 5th International Conference on Computer Graphics, Virtual Reality, Visualisation and Interaction in Africa, pp. 197–204 (2007)
2. Doran, J., Parberry, I.: Controlled procedural terrain generation using software agents. IEEE Transactions on Computational Intelligence and AI in Games 2(2), 111–119 (2010)
3. Ebert, D.S., Musgrave, F.K., Peachey, D., Perlin, K., Worley, S.: Texture and Modeling: A Procedural Approach, 3rd edn. Morgan Kaufmann (2003)
4. Frade, M., de Vega, F.F., Cotta, C.: Modelling video games' landscapes by means of genetic terrain programming: A new approach for improving users' experience. In: Applications of Evolutionary Computing, pp. 485–490 (2008)
5. Frade, M., de Vega, F.F., Cotta, C.: Evolution of artificial terrains for video games based on accessibility. Applications of Evolutionary Computation pp. 90–99 (2010)
6. Frade, M., de Vega, F.F., Cotta, C.: Automatic evolution of programs for procedural generation of terrains for video games. Soft Computing 16(11), 1893–1914 (2012)

7. Kamal, K.R., Uddin, Y.S.: Parametrically controlled terrain generation. In: Proceedings of the 5th International Conference on Computer Graphics and Interactive Techniques in Australia and Southeast Asia, pp. 17–23 (2007)
8. Lechner, T., Ren, P., Watson, B., Brozefski, C., Wilenski, U.: Procedural modeling of urban land use. In: ACM SIGGRAPH 2006 Research posters, p. 135. ACM (2006)
9. Lechner, T., Watson, B., Wilensky, U.: Procedural city modeling. In: Proceedings of the 1st Midwestern Graphics Conference (2003)
10. Mandelbrot, B.B.: The Fractal Geometry of Nature. W.H. Freeman (1982)
11. Musgrave, F.K., Kolb, C.E., Mace, R.S.: The synthesis and rendering of eroded fractal terrains. In: Proceedings of SIGGRAPH 1989, pp. 41–50 (1989)
12. Poli, R., Langdon, W.B., McPhee, N.F.: A Field Guide to Genetic Programming (2008). http://www.gp-field-guide.org.uk
13. Schneider, J., Boldte, T., Westermann, R.: Real-time editing, synthesis, and rendering of infinite landscapes on GPUs. In: Proceedings of the International Symposium on Vision, Modeling and Visualization, pp. 145–152 (2006)
14. Smelik, R.M., De Kraker, K.J., Tutenel, T., Bidarra, R., Groenewegen, S.A.: A survey of procedural methods for terrain modelling. In: Proceedings of the CASA Workshop on 3D Advanced Media in Gaming and Simulation (3AMIGAS) (2009)
15. Togelius, J., Preuss, M., Yannakakis, G.N.: Towards multiobjective procedural map generation. In: Proceedings of the 2010 Workshop on Procedural Content Generation in Games (2010)
16. Zhou, H., Sun, J., Turk, G., Rehg, J.M.: Terrain synthesis from digital elevation models. IEEE Transactions on Visualization and Computer Graphics 13(4), 834–848 (2007)

Chapter 5
Grammars and L-systems with applications to vegetation and levels

Julian Togelius, Noor Shaker, and Joris Dormans

Abstract Grammars are fundamental structures in computer science that also have many applications in procedural content generation. This chapter starts by describing a classic type of grammar, the L-system, and its application to generating plants of various types. It then describes how rules and axioms for L-systems can be created through search-based methods. But grammars are not only useful for plants. Two longer examples discuss the generation of action-adventure levels through graph grammars, and the generation of *Super Mario Bros.* levels through grammatical evolution.

5.1 Plants are everywhere

In the previous chapter we discussed generating terrain. Almost as ubiquitous as terrain itself is vegetation of some form: grass, trees, bushes, and other such plants that populate a landscape. Procedurally generating vegetation is a great fit: we need to create a huge number of artefacts (there are many trees in the forest and many blades of grass in the lawn) that are similar to each other, recognisable, but also slightly different from each other. Just copy-pasting trees won't cut it,[1] because players will quickly spot that every tree is identical. An easy aspect of generating vegetation is that, in most games, it is of little functional significance, meaning that a botched plant will not make the game unplayable, just look a bit weird.

And in fact, vegetation is one of the success stories of PCG. Many games use procedurally generated vegetation, and there are many software frameworks available. For example, the *SpeedTree* middleware has been used in dozens of AAA games.

One of the simplest and best ways to generate a tree or bush is to use a particular form of *formal grammar* called an *L-system*, and interpret its results as drawing instructions. This fact is intimately connected to the "self-similar" nature of plants,

---

[1] In William Gibson's *Neuromancer*, one of the main characters is busy copy-pasting trees in one of the early chapters; Gibson seems not to have anticipated PCG.

N. Shaker et al., *Procedural Content Generation in Games*, Computational Synthesis and Creative Systems, DOI 10.1007/978-3-319-42716-4_5

i.e. that the same structures can be found on both micro and macro levels. For an example of this, take a look at a branch of a fern, and see how the shape of the branch repeats in each sub-branch, and then in each branch of the sub-branch. Or look at a Romanesco broccoli, which consists of cones on top of cones on top of cones, etc. (see Figure 5.1). As we will see, L-systems are naturally suited to reproducing such self-similarity.

Fig. 5.1: Romanesco broccoli. Note the self-similarity. (Photo credit: Jon Sullivan)

In this chapter, we will introduce formal grammars in general, L-systems in particular and how to use a graphical interpretation of L-systems to generate plants. We will also give examples of how L-systems can be used as a representation in search-based PCG, allowing you to evolve plants. However, it turns out that plants are not the only thing for which formal grammars are useful. In the rest of the chapter, we will explain how grammar-based systems can be used to generate quests and dungeon-like environments for adventure games such as *Zelda*, and levels for platform games such as *Super Mario Bros.*

5.2 Grammars

A (formal) *grammar* is a set of *production rules* for rewriting strings, i.e. turning one string into another. Each rule is of the form (symbol(s)) $\rightarrow$ (other symbol(s)). Here are some example production rules:

1. $A \rightarrow AB$
2. $B \rightarrow b$

Using a grammar is as simple as going through a string, and each time a symbol or sequence of symbols that occurs in the left-hand side (LHS) of a rule is found, those symbols are replaced by the right-hand side (RHS) of that rule. For example, if the initial string is *A*, in the first rewriting step the *A* would be replaced by *AB* by rule 1, and the resulting string will be *AB*. In the second rewriting step, the *A* would again be transformed to *AB* and the *B* would be transformed to *b* using rule 2, resulting in the string *ABb*. The third step yields the string '*ABbb* and so on. A convention in grammars is that upper-case characters are nonterminal symbols, which are on the LHS of rules and therefore rewritten further, whereas lower-case characters are terminal symbols which are not rewritten further.

Formal grammars were originally introduced in the 1950s by the linguist Noam Chomsky as a way to model natural language [3]. However, they have since found widespread application in computer science, since many computer science problems can be cast in terms of generating and understanding strings in a formal language. Many results in theoretical computer science and complexity theory are therefore expressed using grammar formalisms. There is a rich taxonomy of grammars which we can only hint at here.[2] Two key distinctions that are relevant for the application of grammars in procedural content generation are whether the grammars are deterministic, and the order in which they are expanded.

Deterministic grammars have exactly one rule that applies to each symbol or sequence of symbols, so that for a given string, it is completely unambiguous which rules to use to rewrite it. In nondeterministic grammars, several rules could apply to a given string, yielding different possible results of a given rewriting step. So, how would you decide which rule to use? One way is to simply choose randomly. In such cases, the grammar might even include probabilities for choosing each rule. Another way is to use some parameters for deciding which way to expand the grammar—we will see an example of this in the section on grammatical evolution towards the end of the chapter.

5.3 L-systems

The other distinction of interest here is in which order the rewriting is done. *Sequential* rewriting goes through the string from left to right and rewrites the string as it is reading it; if a production rule is applied to a symbol, the result of that rule is written into the very same string before the next symbol is considered. In *parallel* rewriting, on the other hand, all the rewriting is done at the same time. Practically, this is implemented as the insertion of a new string at a separate memory location containing only the effects of applying the rules, while the original string is left unchanged. Sometimes, the difference between parallel and sequential rewriting can be major.

L-systems are a class of grammars whose defining feature is parallel rewriting, and which was introduced by the biologist Aristid Lindenmayer in 1968 explicitly to model the growth of organic systems such as plants and algae [9]. The following is a simple L-system defined by Lindenmayer to model yeast growth:

---

[2] For a detailed treatment of formal grammars, and their application to domains other than language, see [16].

1. $A \rightarrow AB$
2. $B \rightarrow A$

Starting with the axiom A (in L-systems the seed strings are called axioms) the first few expansions look as follows:

1. A
2. AB
3. ABA
4. ABAAB
5. ABAABABA
6. ABAABABAABAAB
7. ABAABABAABAABABAABABA
8. ABAABABAABAABABAABABAABAABABAABAAB

There are several interesting things about this sequence. One is the obvious regularity, which is more complex than simply repeating the same string over and over, and certainly seems more complex than is warranted by the apparent simplicity of the system that generates it. But also note that the rate of growth of the strings in each iteration is increasing. In fact, the length of the strings is a Fibonacci sequence: 1 2 3 5 8 13 21 34 55 89... This can be explained by the fact that the string of step n is a concatenation of the string of step $n-1$ and the string of step $n-2$.

Clearly, even simple L-systems have the capacity to give rise to highly complex yet regular results. This seems like an ideal fit for PCG. But how can we move beyond simple strings?

5.3.1 Graphical interpretation of L-systems

One way of using the power of L-systems to generate 2D (and 3D) artefacts is to interpret the generated strings as instructions for a turtle in *turtle graphics*. Think of the turtle as moving across a plane holding a pencil, and simply drawing a line that traces its path. We can give commands to the turtle to move forwards, or to turn left or right. For example, we could use the following key to interpret the generated strings:

- F: move forward a certain distance (e.g. 10 pixels)
- +: turn left 90 degrees
- -: turn right 90 degrees

Such an interpretation can be used in conjunction with a simple L-system to give some rather remarkable results. Consider the following system, consisting only of one rule:

1. $F \rightarrow F + F - F - F + F$

Starting this system with the axiom F, it would expand into $F + F - F - F + F$ and then into $F + F - F - F + F + F + F - F - F + F - F + F - F - F + F - F + F - F - F + F + F + F - F - F + F$ etc. Interpreting these strings as turtle graphics instructions, we get the sequence of rapidly complexifying pyramid-like structures shown in Figure 5.2, known as the Koch curve.

Fig. 5.2: Koch curve generated by the L-system $F \rightarrow F + F - F - F + F$ after 0, 1, 2 and 3 expansions

5.3.2 Bracketed L-systems

While interpreting L-system-generated strings as turtle instructions allows us to draw complex fractal shapes, we are fundamentally limited by the constraint that the figures must be drawable in one continuous line—the whole shape must be drawn "without lifting the pencil". However, many interesting shapes cannot be drawn this way. For example, plants are branching and require you to finish drawing a branch before returning to the stem to draw the next line. For this purpose, *bracketed* L-systems were invented. These L-systems have two extra symbols, [and], which behave like any other symbols when rewriting the strings, but act as "push" and "pop" commands to a stack when interpreting the string graphically. (The stack is simply a first-in, last-out list.) Specifically, [saves the current position and orientation of the turtle onto the stack, and] retrieves the last saved position from the stack and resets the turtle to that position—in effect, the turtle "jumps back" to a position it has previously been at.

Bracketed L-systems can be used to generate surprisingly plant-like structures. Consider the L-system defined by the single rule $F \rightarrow F[-F]F[+F][F]$. This is interpreted as above, except that the turning angles are only 30 degrees rather than 90 degrees as in the previous example. Figure 5.3 shows the graphical interpretation of the L-system after 1, 2, 3 and 4 rewrites starting from the single symbol F. Minor variations of the rule in this system generate different but still plant-like structures, and the general principle can easily be extended to three dimensions by introducing symbols that represent rotation along the axis of drawing. For a multitude of beautiful examples of plants generated by L-systems see the book *The Algorithmic Beauty of Plants* by Prusinkiewicz and Lindenmayer [14].

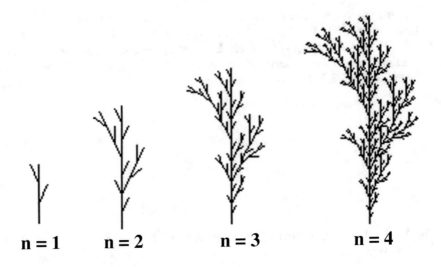

$$n = 1 \qquad n = 2 \qquad n = 3 \qquad n = 4$$

Fig. 5.3: Four rewrites of the bracketed L-system $F \rightarrow F[-F]F[+F][F]$

5.4 Evolving L-systems

As with other parameterized representations for procedural content, L-system expansions can be used as genotype-to-phenotype mappings in search-based PCG. An early paper by Ochoa presents a method for evolving L-systems to attain particular 2D shapes [10]. She restricts herself to L-systems with the simple alphabet used above ($F + -[]$), the axiom F, and a single rule with the LHS F. The genotype is the RHS of the single rule. Ochoa used a canonical genetic algorithm with crossover and mutation together with a combination of several evaluation functions. The fitness functions relate to the shape of the phenotype: the height ("phototropism"), symmetry, exposed surface area ("light-gathering ability"), structural stability, and proportion of branching points. By varying the contributions of each fitness function, she showed that it is possible to control the type of the plants generated with some precision. Figure 5.4 shows some examples of plants evolved with a combination of fitness functions, and Figure 5.5 shows some examples of organism-like structures evolved with the same representation but a fitness function favouring symmetry.

5.5 Generating missions and spaces with grammars

A game level is not a singular construction, but rather a combination of two interacting structures: a mission and a space [4]. A mission describes the things a player can or must do to complete a level, while the space describes the geometric layout of the

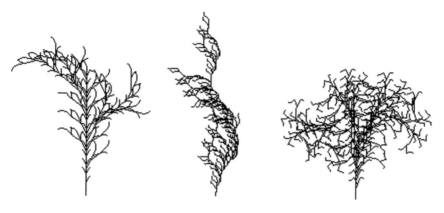

Fig. 5.4: Some evolved L-system plants. Adapted from [10]

Fig. 5.5: Some L-system structures evolved for symmetry. Adapted from [10]

environment. Both mission and space have their own structural qualities. For missions it is important to keep track of flow, pacing and causality, while for the space connectedness, distance and sign posting are critical dimensions. To successfully generate levels that feel consistent and coherent, it is important to use techniques that can generate each structure in a way that strengthens its individual qualities while making sure that the two structures are interrelated and work together. This section discusses how different types of generative or transformative grammars can be used to achieve this.

5.5.1 Graph grammars

Generative grammars typically operate on strings, but they are not restricted to that type of representation. Grammars can be used to generate many different types of structures: graphs, tile maps, two- or three-dimensional shapes, and so on. In this section and the following section, we will explore how grammars can be used to generate graphs and tile maps. These structures are useful ways to represent game missions and game spaces that combine to make game levels.

Graphs are more useful than strings to represent missions and spaces for games, especially when these missions and spaces need to have a certain level of sophisti-

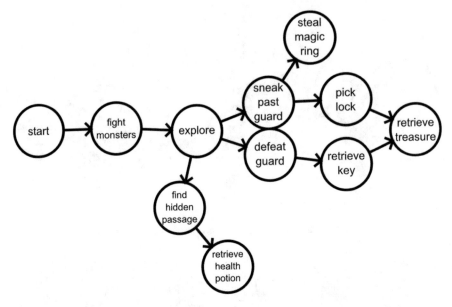

Fig. 5.6: A mission structure with two paths

cation. For example, a completely linear mission (which might be represented by a string) might be suitable for simple and linear games, but for explorative adventure games such as RPG dungeons you would want missions to contain lock and key puzzles, bonus objectives, and possibly multiple paths to lead to the level goal. Graphs can express this type of structure more easily. For example, Figure 5.6 contains a mission that can be solved in two different ways.

Graph grammars work quite similarly to string grammars; graph grammar rules also have a left-hand part that identifies a particular graph construction that can be replaced by one of the constructions in the right-hand part of the rule. However, to make the transformation, it is important to identify each node on the left-hand part individually and to match them with individual nodes in each right-hand part. Figure 5.7 represents a graph grammar rule and uses numbers to identify each individual node. When using this rule to transform a graph, the following five steps are performed (as illustrated by Figure 5.8)[3] [15]:

1. Find a subgraph in the target graph that matches the left-hand part of the rule and mark that subgraph by copying the identifiers of the nodes.
2. Remove all edges between the marked nodes.
3. Transform the graph by transforming marked nodes into their corresponding nodes on the right-hand side, adding a node for each node on the right-hand

---

[3] In simple graph transformations there is no need to identify and transform individual edges in the same way as nodes are identified and transformed. However, a more sophisticated implementation that requires edges to be transformed rather than removed and added for each transformation can be realised by identifying and replacing edges in the same way as nodes.

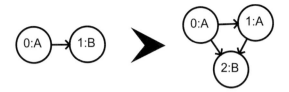

Fig. 5.7: A graph grammar rule

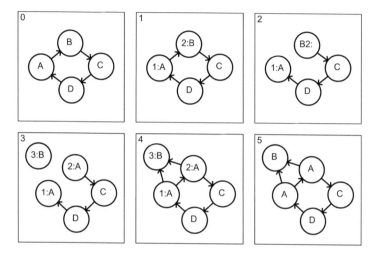

Fig. 5.8: Graph grammar transformation

side that has no match in the target graph, and removing any nodes that have no corresponding node on the right-hand side.[4]

4. Copy the edges as specified by the right-hand side.
5. Remove all marks.

5.5.2 Using graph grammars to generate missions

To generate a simple mission using graph grammars, it is best to start by defining the alphabet the grammar is designed to work with. In this case the alphabet consists of the following nodes and edges:

- Start (node marked S): the start symbol from which the grammar generates a mission (the axiom).
- Entrance (nodes marked e): the starting place of the player.

---

[4] The removal of nodes only works when the node to be removed is only connected to nodes that have been marked. This is something to take into account when designing graph grammar rules.

- Tasks (nodes marked t): arbitrary, unspecified tasks (here be monsters!).
- Goals (nodes marked g): a task that finishes the level when successfully completed.
- Locks (nodes marked l): a task that requires a key to perform successfully.
- Keys (nodes marked k).
- Non-terminal task nodes (nodes marked T).
- Normal edges (represented as solid arrows) connecting nodes and identifying which task follows which.
- Unlock edges (represented as solid arrows marked with a dash) connecting keys to locks.

With this alphabet we can construct rules that generate missions. For example, the rules in Figure 5.9 were used to generate the sample missions in Figure 5.10.[5]

One thing you might notice from studying these rules is that graph grammars can be hard to control. In the case of the rule set represented in Figure 5.9, the number of tasks generated (by the application of the "add task" rule) can be as low as one and has no upper limit. As soon as the Start node is removed from the graph, the number of tasks no longer grows. One way to get a better grip on the generated structures is not to apply rules indiscriminately, but to specify a sequence of rules so that each rule in the sequence is applied once to one possible location in the graph. For example, if we split up the "add task" rule from Figure 5.9 into two rules (see Figure 5.11), the missions in Figure 5.12 are generated by applying the following sequence of rules:[6]

- start rule (x1),
- add task (x6),
- add boss (x1),
- define task (x6),
- move lock (x5).

5.5.3 Breaking the process down into multiple generation steps

So far, the graph grammars are relatively simple. However, to generate anything resembling the complexity of the mission in Figure 5.6, many more rules are required. Designing the grammars to achieve such results takes practice and patience. A key strategy for designing successful grammars is to break the process down into

---

[5] The rules use a special wildcard node (marked with a *) to indicate a match with any node. Wildcards on the right-hand side of a rule never change the corresponding node in the graph being transformed. An alternative to these wildcards is to allow rules to have edges without origin or target node.

[6] Obviously, the sequence of rules might be generated by a string grammar.

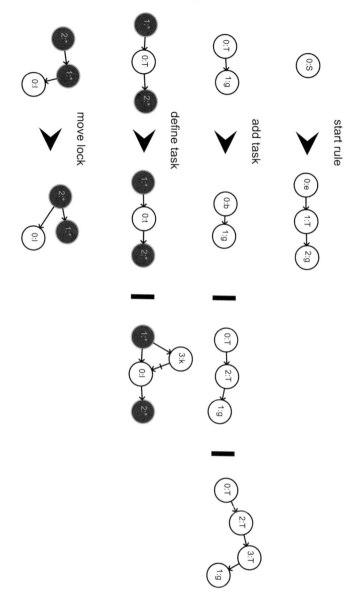

Fig. 5.9: Mission rules

multiple steps. Trying to generate everything at once using only one grammar is a daunting task, and next to impossible to debug and maintain.[7]

---

[7] Breaking the generation down into multiple steps is in line with the approach to software engineering and code generation suggested by model-driven engineering. When done right, this ap-

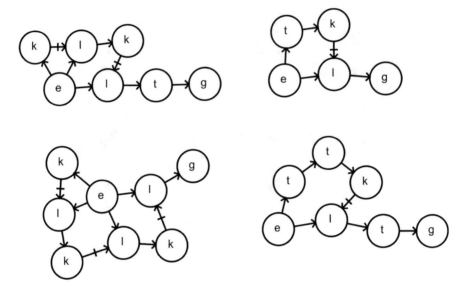

Fig. 5.10: Generated missions

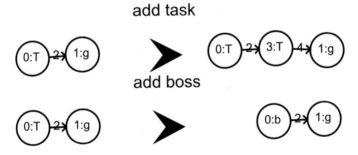

Fig. 5.11: Two new rules to replace the old "add task"

When breaking the generation process down into multiple steps, it is useful to think of each step as a simulation of the design process. One step might generate the overall specifications of the mission, while the next might flesh out those specifications. In game design, a successful design strategy is to start from a random set of requirements and use your creativity to shape that random collection into a coherent whole. Following a similar approach for breaking down the generation procedure and designing individual grammars yields good results. In particular, designing one simple step to create a highly randomised graph and using a second step to restructure that graph into something that makes sense from the game's perspective is an effective strategy to create expressive generation procedures [6].

---

proach leads to a flexible generation process that allows you to generate spaces from missions or vice versa, and creates opportunities to design generic, reusable generation steps [1, 5].

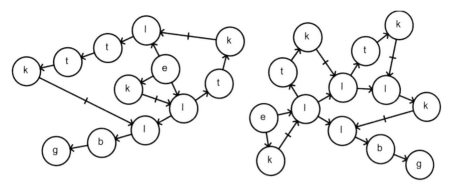

Fig. 5.12: Missions generated from the same sequence of rules

For example, we can use a single step to generate a mission of a specified length and randomly choose between locks, keys and other tasks to fill in the spaces between the entrance and the goal. In this case we also make sure that the first task is always a key and the last task is always a lock. Figure 5.13 and Figure 5.14 represent the rules and a sample mission built using those rules. Note that although locks and keys are placed, no relationship between them is established.

The next step is to extract lock and key relationships. Based on the spread of the locks and keys over the tasks, multiple keys can be assigned to a single lock, and vice versa. This would represent multiple levers that need to be activated to open a single door, or a special weapon that can be used multiple times to get past a special type of barrier. Figure 5.15 represents the rules to add these relationships, and Figure 5.16 is a sample configuration created from the sample set in Figure 5.14.

Subsequent steps could include the movement of locks through the graph (as we have seen in the example above), generating more details of the nature of the locks and keys, or adding tasks of a different type. One of the advantages of using these two steps is that two relatively simple grammars can create a large variety of different relationships (two keys to a single lock, or keys that are reused). Getting the same level of variation using explicit rules that create X number of keys to a single lock would require many more rules, which are much harder to maintain. In addition, the second step can also be executed on graphs that have been built to different specifications. For example, the same rules can be used to create lock and key relationships for a dungeon that has two separate paths (see Figure 5.17).

5.5.4 Generating spaces to accommodate a mission

Having a representation of a mission itself is only one step towards the generation of levels for a game. Missions need to be transformed into spaces that the player can actually traverse. Transforming from mission to space is one of the hardest steps in

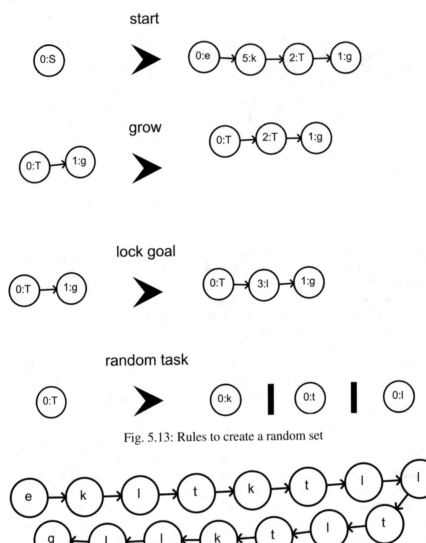

Fig. 5.13: Rules to create a random set

Fig. 5.14: Sample random set

this process. The problem comes down to generating two different, independent but linked structures: an abstract mission that details the things a player needs to do, and a concrete space that creates the world where the player can do these things. Below are three strategies to deal with the problem of generating the two structures:

1. Transform from mission to space. The transition from Figure 5.14 to Figure 5.16 reflects the gradual transition from an abstract mission to a more concrete repre-

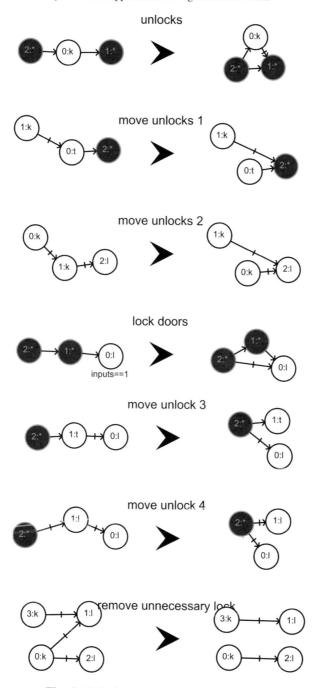

Fig. 5.15: Rules to add lock and key relationships

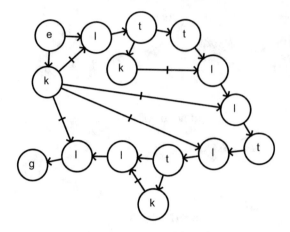

Fig. 5.16: Generated lock and key relationships

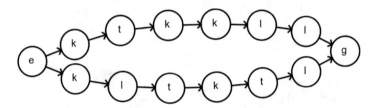

Fig. 5.17: Two paths to a single goal

sentation of a game space, although, in this case, the game space is still highly abstract. However, by using automatic graph layout algorithms and sampling the results into a tile map, it is possible to generate usable level geometry. This approach works well for games such as action-adventure games or games with a strong narrative, where mission coherence and pacing is important. The disadvantage of this approach is that the difficulty of going from mission to space is most pronounced.

2. Transform a mission into a set of instructions to build a space. Instead of directly transforming a mission structure into a space, it is possible to transform the mission into a set of building instructions that can be used to build a space to match the requirements. This approach has the advantage that the transition from graphs to tiles or shapes is much easier. It also comes at a cost: it is very difficult to generate spaces that have multiple paths leading to the same goal or location. So this approach works best for very linear games like platformers or certain story-driven games.

3. Build level geometry and distill a more abstract representation of the game space from which to generate the missions. This approach inverts the problem by generating level geometry first and then setting up missions for that geometry. This can be done by generating geometry using cellular automata, grammars, evolu-

tion, or any other technique, then analysing the geometry to create an abstract graph representation of the same space, which can be transformed into suitable mission structures. This approach works well for strategic games, levels that take place in locations that require some consistent architecture (such as castles, dwarf fortresses, police stations, or space ships) and for levels that the player is going to visit multiple times. The downside of this approach is that it is critical that the geometry is generated with enough mission potential (are there doors to be locked, bottlenecks suitable for traps, and so on?). There is also less control over the mission than with the other two approaches.

When choosing between these strategies, or when trying to come up with another strategy, it is important to think like a designer. The most effective way of generating levels using a multistep process and different representations of missions and spaces is to model the real design process. Ask yourself, how would you go about designing a level by hand? Would you start by listing mission goals, or by sketching out a map? What sort of changes do you make and can those changes be captured by transformational grammars?

5.5.5 Extended example: 'Dules

An extended example following the third strategy concludes this section. This example details part of the PCG for the game 'Dules, which is currently in development. In this game, players control futuristic combat vehicles (tanks, hovercraft, and so on) in a post-apocalyptic, alien-infested world. The players can choose missions from a world map, after which the game generates an environment to match the location on the map and sets up a mission based on the affordances of the environment and specifications dictated by the current game state (who controls the environment, is the player trying to take over or defending from alien incursion, and so on).

The content generation of 'Dules makes use of transformation grammars that operate on strings, graphs, and tiles. Tile grammars are very simple. They also consist of rules with one left and one or more right hands where the left hand can be replaced by one of the right-hand constructions. Like graph grammars, the tile grammars used in 'Dules can work with wildcards to indicate that certain tiles can be ignored. In contrast to string and graph grammars, tile grammars cannot change the number of tiles. In addition, tile grammars can be made to stack tiles onto each other instead of replacing them.

The procedural content generation procedure roughly follows the steps outlined Figure 5.18. The tile-based world map is taken as input (1), and the particular location is selected (2). Based on the presence of particular tiles indicating vegetation, elevation, buildings, and so on, a combination of tile grammars and cellular automata are used to create the terrain (3-7). The terrain is analysed and transformed into an abstract representation (8). At the same time, mission specifications are generated using a string grammar (9), and these are used as building instructions to plot

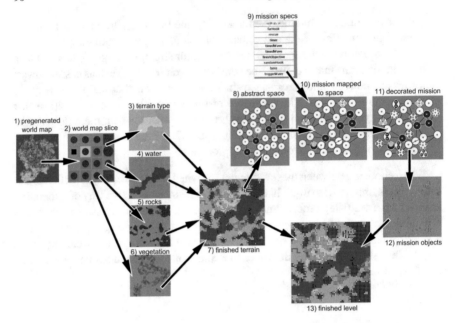

Fig. 5.18: The generation steps to create a level for *'Dules*

a mission onto the space graph (10).[8] Finally, some extra enemies are added to the mission (11), and all the mission-specific game objects are placed onto the same tile map (12) and combined with the terrain to create the complete mission (13).

Almost all steps in the process are handled by grammars. Tile grammars are used to generate the terrain; tile grammars are even used to specify different cellular automata. String grammars are used to create the mission specification and graph grammars are used to create the mission itself. The translation of the terrain into the space graph is done using a specialised algorithm that distinguishes between walkable terrain, impassible terrain, and bodies of water. Each node in (8) represents around 100 tiles, and a reference between the node and the tiles is kept so that the game objects may be placed in the right area during (12).

5.6 Grammatical evolution for *Infinite Mario Bros.* level generation

Grammatical evolution (GE) is an evolutionary algorithm based on genetic programming (GP) [12]. The main difference between GE and GP is in the genome

---

[8] In this case certain graph nodes are depicted as containing other nodes. This is just a depiction: for the implementation and the grammars, containment is simply a special type of edge that is rendered differently.

representation; while a tree-based structure is used in GP, GE relies on a linear genome representation. Like general genetic algorithms (GAs), GE applies fitness calculations for every individual and then applies genetic operators to produce the next generation.

The population of the evolutionary algorithm consists of variable-length integer vectors, initialised randomly. The syntax of possible solutions is specified through a context-free grammar. GE uses the grammar to guide the construction of the phenotype output. The context-free grammar employed by GE is usually written in Backus-Naur form (BNF). Because of the use of a grammar, GE is capable of generating anything that can be described as a set of rules such as mathematical formulas [18], programming code, game levels [17] and physical and architectural designs [2, 13]. GE has been used intensively for automatic design [8, 2, 13, 7, 11], a domain where it has been shown to have a number of advantages over more traditional optimisation methods.

5.6.1 Backus-Naur form

Backus-Naur form (BNF) is common format for expressing grammars. A BNF grammar $G = \{N, T, P, S\}$ consists of terminals, T, non-terminals, N, production rules, P, and a start symbol, S. As in any grammar, non-terminals can be expanded into one or more terminals and non-terminals through applying the production rules. An example BNF to generate valid mathematical expressions is given in Figure 5.19.

```
(1) <exp> ::= <exp> <op> <exp>
          | ( <exp> <op> <exp> )
          | <var>
(2) <op> :: = + | - | * | /
(3) <var> ::= X
```

Fig. 5.19: Illustrative grammar for generating mathematical expressions

Each chromosome in GE is a vector of codons. Each codon is an integer used to select a production rule from the BNF grammar in the genotype-to-phenotype mapping. A complete program is generated by selecting production rules from the grammar until all non-terminals are replaced. The resulting string is evaluated according to a fitness function to give a score to the genome. To better understand the genotype-to-phenotype mapping, we will give a brief example.

Consider the grammar in Figure 5.19 and the individual genotype integer string $(4, 5, 8, 11)$. We begin the processing of the mapping from the start symbol $< exp >$. There are three possible productions; to decide which production to choose, we use the first value in the input genome and apply the mapping function $4\%3 = 1$, where 3 is the number of possible productions. The result from this operation indicates that the second production should be chosen, and $< exp >$ is replaced with $(<$

$exp><op><exp>)$. The mapping continues by using the next integer with the first unmapped symbol in the mapping string; the mapping string then becomes $(<var><op><exp>)$ through the formula $5\%3 = 2$. At this step $<var>$ has only one possible outcome and there is no choice to be made, hence, X is inserted without reading any number from the genome. The expression becomes $(X<op><exp>)$. Continuing to read the codon values from the example individual's genome, $<op>$ is mapped to $+$ and $<exp>$ is mapped to X through the two formulas, $8\%4 = 0$ and $11\%3 = 2$, respectively. This results in the expansion $(X+X)$.

During the mapping process, it is possible for individuals to run out of genes, in which case GE either declares the individual invalid by assigning it a penalty fitness value, or it wraps around and reuses the genes.

5.6.2 Grammatical evolution level generator

Shaker et al. [17] used grammatical evolution to generate content for *Infinite Mario Bros*. It has a number of advantages for this task: it provides a simple way of describing the structure of the levels; it enables an open-ended structure where the design and model size are not known a priori; it enables the design of aesthetically pleasing levels by exploring a wide space of possibilities since the exploratory process is not constrained or biased by imagination or known solutions; it allows easy incorporation of domain knowledge through its underlying grammatical representation, which permits level designers to maintain greater control of the output; finally, it is easily generalised to different types of games.

The following section summarises the work of Shaker et al. [17]. We start by presenting the design grammar used by GE to specify the structure of IMB levels; after that we present how GE was employed to evolve playable levels for the game.

5.6.2.1 Design grammar for content representation

As mentioned earlier, GE uses a design grammar (DG), written in BNF, to represent solutions (in our case a level design). Several methods can be followed to specify the structure of the levels in a design grammar, but since the grammar employed by GE is a context-free grammar, this limits the possible solutions available. To accommodate this constraint, and to keep the grammar as simple as possible, the work here adds game elements to the 2D level array regardless of the positioning of other elements. With this solution, however, arise a number of conflicts in level design that must be resolved. The next section will discuss this conflict-resolution issue and a solution in detail.

The internal representation of the levels in IMB is a two-dimensional array of objects, such as brick blocks, coins and enemies. The levels are generated by placing a number of chunks in the two-dimensional level map. The list of chunks that was considered includes platforms, gaps, stairs, piranha plants, bill blasters, boxes

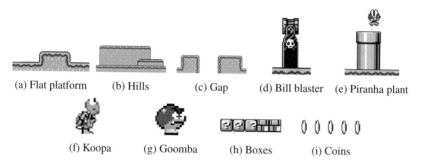

Fig. 5.20: The chunks used to construct *Infinite Mario Bros.* levels

(blocks and brick blocks), coins, goombas and koopas. Each of these chunks has a distinguishable geometry and properties. Figure 5.20 presents the different chunks that collectively constitute a level. The level initially contains a flat platform that spans the whole x-axis; this explains the need to define a gap as one of the chunks.

A design grammar was specified that takes into account the different chunks. In order to allow more variations in the design, platforms and hills of different types were considered such as a blank platform/hill, a platform/hill with a bill blaster, and a platform/hill with a piranha plant.

Variations in enemy placements were achieved by (1) constructing the physical structure of the level, (2) calculating the possible positions at which an enemy can be placed (this includes all positions where a platform was generated) and (3) placing each generated enemy in one of the possible positions.

The design grammar constructed can be seen in Figure 5.21. A level is constructed by placing a number of chunks, each assigned two or more properties; the x and y parameters specify the coordinates of the chunk starting position in the 2D level array and are limited to the ranges [5, 95] and [3, 5], respectively. These ranges are constrained by the dimensions of the level map. The first and last five blocks in the x dimension are reserved for the starting platform and the ending gate, while the y values have been constrained in a way that ensures playability (the existence of a path from the start to the end position) by placing all items in areas reachable by jumping. The w_g parameter specifies the width of gaps that ensures the ability to reach the other edge, w stands for the width of a platform or a hill, w_c defines the number of coins, and h indicates the height of tubes, piranha plants, or the bill blaster. This height is also constrained to the range [3, 4], ensuring that tubes and bill blasters can be jumped over.

5.6.2.2 Conflict resolution and content quality

There are a number of conflicts inherent in the design grammar. Each generated chunk can be assigned any x and y values from the ranges [5, 95] and [3, 5], respectively, depending on the genotype. This means it is likely there will be an over-

```
<level> ::= <chunks>  <enemy>
<chunks> ::= <chunk> |<chunk> <chunks>
<chunk> ::= gap(<x>,<y>, <Wg>,<Wbefore>,<Wafter>)
  | platform(<x>,<y>,<w>)
  | hill(<x>,<y>,<w>)
  | blaster_hill(<x>,<y>,<h>,<Wbefore>,<Wafter>)
  | tube_hill(<x>,<y>,<h>,<Wbefore>,<Wafter>)
  | coin(<x>,<y>,<Wc>)
  | blaster(<x>,<y>,<h>,<Wbefore>,<Wafter>)
  | tube(<x>,<y>,<h>,<Wbefore>,<Wafter>)
  | <boxes>

<boxes> ::=  <box_type> (<x>,<y>)² | ...
  | <box_type> (<x>,<y>)⁶

<box_type> ::= blockcoin | blockpowerup
  | brickcoin | brickempty

<enemy> ::= (koopa | goomba)(<pos>) ² | ...
  | (koopa | goomba)(<pos>) ¹⁰
<x> ::= [5..95]
<y> ::= [3..5]
<Wg> ::= [2..5]
<Wbefore> ::= [2..5]
<Wafter> ::= [2..5]
<w> ::= [2..6]
<Wc> ::= [2..6]
<h> ::= [3..4]
<pos> ::= [0..100000]
```

Fig. 5.21: The design grammar employed to specify the design of the level. The superscripts (2, 6 and 10) are shortcuts specifying the number of repetition

lap between the coordinates of the generated chunks. For example, $hill(65,4,5)$ $hill(25,4,4)$ $blaster\_hill(67,4,4,4,3)$ $coin(22,4,6)$ $platform(61,4,4)$ is a phenotype that has been generated by the grammar and contains a number of conflicts: e.g. $hill(65,4,5)$ and $blaster\_hill(67,4,4,4,3)$ were assigned the same y value, and overlapping x values; another conflict occurs between $hill(25,4,4)$ and $coin(22,4,6)$ as the two chunks also overlap.

   To resolve these conflicts, a manually defined priority value is assigned to each chunk. Hills with bill blasters or piranha plants are given the highest priority, followed by blank hills, platforms with enemies (bill blasters or piranha plants) come next then blank platforms and finally come coins and blocks with the lowest priority. After generating a genotype (with possible conflicts), a post-processing step is applied in which the chunks are arranged in descending order according to their priorities, coordinates and type. The resulting ordered phenotype is then scanned and whenever two overlapping chunks are detected, the one with the higher priority value is maintained and the other is removed. Nevertheless, to allow more diver-

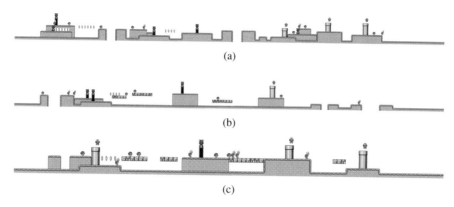

Fig. 5.22: Example levels generated by the GE-generator using the design grammar in Figure 5.21

sity, some of the chunks are allowed to overlap such as hills of different heights (Figure 5.20b), and coins or boxes with hills. Without this refinement, most levels would look rather flat and uninteresting.

A relatively simple fitness function is used to measure content quality. The main objective of the fitness function is to allow exploration of the design space by creating levels with an acceptable number of chunks, giving rich design and variability. Thus, the fitness function used is a weighted sum of two normalised measures: the first one, f_p, is the difference between the number of chunks placed in the level and a predefined threshold that specifies the maximum number of chunks that can be placed. The second, f_c, is the number of different conflicting chunks found in the design. Apparently, the two fitness functions partially conflict since optimising f_p by placing more chunks implicitly increases the chance of creating conflicting chunks (f_c). Some example levels generated are presented in Figure 5.22.

5.7 Lab exercise: Create plants with L-systems

In this lab exercise, you will implement a simple bracketed L-system to generate plants. Use an L-system to generate your plants and a turtle graphics program to draw them. You will be given a software package that contains three main classes: *LSystem*, *State* and *Canvas*. Your main work will be to implement the two main methods in the *LSystem* class:

```
public void expand(int depth)
public void interpret(String expression)
```

The L-system has an alphabet, axioms, production rules, a starting point, a starting angle, a turning angle and a length for each step. The expand method is used to

Fig. 5.23: Example trees generated with an L-system using different instantiation parameters

expand the axiom of the L-system a number of times specified by the *depth* parameter. After expansion, the system processes the expansion and visualises it through the *interpret* method. The result of each step is drawn on the canvas. Since the L-system will be in a number of different states during expansion, a *State* class is defined to represent each state. An instance of this class is made for each state of the L-system and the variables required for defining the state are passed on from the L-system to the state; these include the x and y coordinates, the starting and turning angles and the length of the step. The L-system is visualised by gradually drawing each of its states.

The *State* and the *Canvas* classes are helpers, and therefore there is no need to modify them. The *Canvas* class has the methods required for simple drawing on the canvas and it contains the main method to run your program. In the *main* method, you can instantiate your L-system, define your axiom and production rules and the number of expansions. Figure 5.23 presents example L-systems generated using the following rules: $(F, F, F \rightarrow FF - [-F + F + F] + [+F - F - F])$ (left) and $(F, f, (F \rightarrow FF, f \rightarrow F - [[f] + f] + F[+Ff] - f))$ (right). Note that the rules are written in the form $G = (A, S, P)$, where A is the alphabet, S is the axiom or starting point and P is the set of production rules.

You can use the same software to draw fractal-like forms such as the ones presented in Figure 5.24. Some simple example rules that can be used to create relatively complex shapes are the following: $(F, F + F + F + F, (F + F + F + F \rightarrow F + F + F + F, F \rightarrow F + F - F - FF + F + F - F))$ (left), $(F, F + +F + +F, F \rightarrow F - F + +F - F)$ (middle) and $(F, f, (f \rightarrow F - f - F, F \rightarrow f + F + f))$ (right).

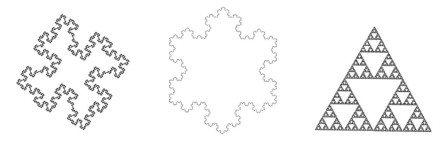

Fig. 5.24: Example fractals generated with an L-system using different production rules

5.8 Summary

Grammars can be useful for creating a number of different types of game content. Perhaps most famously, they can be used to create plant structures; plants generated by grammars are now commonplace in commercial games and game engines. But grammars can also be used to generate levels and physical structures of various kinds and mission structures. Grammars are characterised by expanding an axiom through production rules. The L-system is a simple grammar characterised by simultaneous expansion of symbols, which can generate strings with repeating structure. If the symbols in the string are interpreted as instructions for a movable "pen", the results of the grammar can be interpreted as geometrical patterns. Adding bracketing to a grammar makes it possible to create strings that can be interpreted as branching structures, e.g. trees. Both grammars and rules can be created through search-based methods such as evolution, making automatic grammar design possible. Graph grammars and space grammars extend the basic idea of grammars beyond strings, and can be useful for generating level structures or quest structures.

References

1. Brown, A.: An introduction to model driven architecture (2004).   URL http://www.ibm.com/developerworks/rational/library/3100.html
2. Byrne, J., Fenton, M., Hemberg, E., McDermott, J., O'Neill, M., Shotton, E., Nally, C.: Combining structural analysis and multi-objective criteria for evolutionary architectural design. Applications of Evolutionary Computation pp. 204–213 (2011)
3. Chomsky, N.: Three models for the description of language. IRE Transactions on Information Theory **2**(3), 113–124 (1956)
4. Dormans, J.: Adventures in level design: Generating missions and spaces for action adventure games. In: Proceedings of the Foundations of Digital Games Conference (2010)
5. Dormans, J.: Level design as model transformation: A strategy for automated content generation. In: Proceedings of the Foundations of Digital Games Conference (2011)
6. Dormans, J., Leijnen, S.: Combinatorial and exploratory creativity in procedural content generation. In: Proceedings of the Foundations of Digital Games Conference (2013)

7. Hemberg, M., O'Reilly, U.: Extending grammatical evolution to evolve digital surfaces with Genr8. In: Proceedings of the 7th European Conference on Genetic Programming, pp. 299–308 (2004)
8. Hornby, G., Pollack, J.: The advantages of generative grammatical encodings for physical design. In: Proceedings of the IEEE Congress on Evolutionary Computation, pp. 600–607 (2001)
9. Lindenmayer, A.: Mathematical models for cellular interactions in development I. filaments with one-sided inputs. Journal of Theoretical Biology **18**(3), 280–299 (1968)
10. Ochoa, G.: On genetic algorithms and Lindenmayer systems. In: Parallel Problem Solving from Nature, pp. 335–344. Springer (1998)
11. O'Neill, M., Brabazon, A.: Evolving a logo design using Lindenmayer systems, postscript & grammatical evolution. In: Proceedings of the IEEE Congress on Evolutionary Computation, pp. 3788–3794 (2008)
12. O'Neill, M., Ryan, C.: Grammatical evolution. IEEE Transactions on Evolutionary Computation **5**(4), 349–358 (2001)
13. O'Neill, M., Swafford, J., McDermott, J., Byrne, J., Brabazon, A., Shotton, E., McNally, C., Hemberg, M.: Shape grammars and grammatical evolution for evolutionary design. In: Proceedings of the 11th Conference on Genetic and Evolutionary Computation, pp. 1035–1042 (2009)
14. Prusinkiewicz, P., Lindenmayer, A.: The Algorithmic Beauty of Plants. Springer (1990)
15. Rekers, J., Schürr, A.: A graph grammar approach to graphical parsing. In: Proceedings of the 11th IEEE Symposium on Visual Languages, pp. 195–202 (1995)
16. Rozenberg, G., Salomaa, A. (eds.): Handbook of Formal Languages, vol. 3: Beyond Words. Springer (1997)
17. Shaker, N., Nicolau, M., Yannakakis, G.N., Togelius, J., O'Neill, M.: Evolving levels for Super Mario Bros. using grammatical evolution. In: Proceedings of the IEEE Conference on Computational Intelligence and Games, pp. 304–311 (2012)
18. Tsoulos, I., Lagaris, I.: Solving differential equations with genetic programming. Genetic Programming and Evolvable Machines **7**(1), 33–54 (2006)

Chapter 6
Rules and mechanics

Mark J. Nelson, Julian Togelius, Cameron Browne, and Michael Cook

Abstract Rules are at the core of many games. So how about generating them? This chapter discusses various ways to encode and generate game rules, and occasionally game entities that are strongly tied to rules. The first part discusses ways of generating rules for board games, including *Ludi*, perhaps the most successful example of automatically generated game rules. The second part discusses some more tentative attempts to generate rules for video games, in particular 2D games with graphical logic. Most approaches to generating game rules have used search-based methods such as evolution, but there are also some solver-based approaches.

6.1 Rules of the game

So far in this book, we have seen a large number of methods for generating content for existing games. If you have a game already, that means you can now generate many things for it: maps, levels, terrain, vegetation, weapons, dungeons, racing tracks. But what if you don't already have a game, and want to generate the game itself? What would you generate, and how? At the heart of many types of games is a system of game rules. This chapter will discuss representations for game rules of different kinds, along with methods to generate them, and evaluation functions and constraints that help us judge complete games rather than just isolated content artefacts.

Our main focus here will be on methods for generating interesting, fun, and/or balanced game rules. However, an important perspective that will permeate the chapter is that game rule encodings and evaluation functions can encode game design expertise and style, and thus help us understand game design. By formalising aspects of the game rules, we define a space of possible rules more precisely than could be done through writing about rules in qualitative terms; and by choosing which aspects of the rules to formalise, we define what aspects of the game are interesting to explore and introduce variation in. In this way, each game generator

© Springer International Publishing Switzerland 2016
N. Shaker et al., *Procedural Content Generation in Games*, Computational
Synthesis and Creative Systems, DOI 10.1007/978-3-319-42716-4_6

can be thought of as an executable micro-theory of game design, though often a simplified, and sometimes even a caricatured one [32].

6.2 Encoding game rules

To generate game rules, we need some way of *representing* or *encoding* them in a machine-readable format that some software system can work with.[1] An ambitious starting point for a game encoding might be one that can encode game rules *in general*: an open-ended way to represent any possible game. The game generator would then work on games in this encoding, looking for variants or entirely new games in this space. But such a fully general encoding provides a quite unhelpful starting point. A completely general representation for games cannot say very much that is specific about games at all. Some kinds of games have turns, but some don't. Some games are primarily about graphics and movement, while others take place in an abstract mathematical space. The only fully general encoding of a computer game would be simply a general encoding for all software. Something like "C source code" would suffice, but it produces an extremely *sparse* search space. Although all computer games could in principle be represented in the C programming language, almost all things that can be represented in C's syntax are not in fact games, and indeed many of them are not even working programs, making a generator's job quite difficult.[2]

Instead of having a generator search through the extremely sparse space of all computer programs to find interesting games, a more fruitful starting point is to pick an encoding where the space includes a more dense distribution of things that are games and meet some basic criteria of playability. That way, our generator can spend most of its time attempting to design interesting game variants. Furthermore, it's helpful for game encodings to start with a specific genre. Once we restrict focus to a particular genre, it's possible to abstract meaningful elements common to games in the genre, which the generator can take as given. For example, an encoding for turn-based board games can assume that the game's time advances in alternating discrete turns, that there are pieces on spaces arranged in some configuration, and that play is largely based on moving pieces around. This means the game generator does not have to invent the concept of a "turn", but instead can focus on finding interesting rules for turn-based board games. An encoding for a side-scrolling space shooter would be very different: here the encoding would include continuous time;

---

[1] There are many other uses for machine-readable game rules, such as for use in game-playing AI competitions [12, 6] and in game-design assistants targeted at human game designers [20, 9]. This chapter focuses on encodings for *generating* rules, but multi-use encodings are often desirable.

[2] This is not to say generating games encoded as raw programs would be *impossible*: genetic-programming techniques evolve programs encoded in fairly general representations [29], and applying genetic programming to videogame design could produce interesting results. But the techniques in this chapter focus on higher-level representations, which allow the generators to work on more familiar game-design elements rather than on low-level source code.

entities such as terrain, enemies, physics, lives, and spawn points; and events such as shooting, object collision, and scrolling. Of course, the encoding cannot be *too* narrow: at the limit, an encoding that specifies exactly one game (or only a few) is not very interesting for a game-generation system. The most productive point on the spectrum between complete generality and complete specificity is one of the key tradeoffs in designing an encoding for game generators to use: smaller spaces typically are more dense in playable, interesting candidates, but larger spaces may allow for more interesting variation [27].[3]

In addition to being a more fruitful space for game generators to work in, genre-specific encodings also make it easier to produce *playable* games. Whereas a computer could generate purely abstract rule systems, making interesting games that are playable by humans requires connecting those abstract rules to concrete audiovisual representations [18]. For example, the abstract notion of a "capture" in board games is often represented by physically removing a piece from the board. The idea of "hidden information" in card games is represented by how players hold their cards, and which cards on the table are face up versus face down. Concepts such as "health" can be represented in any number of ways, ranging from numerical display of hitpoints or health percentage on the screen, to more indirect methods such as changing a character's colour, or even varying the music when a player's health drops below a threshold. Matching generated rules to these concrete representations can be a challenging research problem in itself [25], but working with encodings of specific genres allows us to sidestep the issue, by having a standard concrete representation for the genre being considered.

Finally, using a genre-specific encoding provides a first step towards answering a key question: how do we evaluate what constitutes a good set of game rules? Rather than the extremely general question of what makes a good game, we can ask what makes a good *two-player board game*, a good *real-time strategy game*, or a good *first-person shooter*. That lets us take advantage of existing genre-specific design knowledge, which is usually better developed and more amenable to being formalised. Design of new board games may focus on properties such as balance, availability of multiple nontrivial strategies, etc. Criteria for designing a good side-scrolling shooter, meanwhile, may instead focus on the pace of the action, patterns of enemy waves, and the difficulty progression—very different kinds of criteria. When we generate the rules for games using encodings of these well-defined genres, we can use a wide variety of existing design knowledge to made our playability and quality judgements. This allows rule-generating PCG systems to start from the basis of being *domain experts* in a specific genre, to use Khaled et al.'s terms for PCG system roles [14].

The two sections that follow describe game-generator experiments that a number of researchers have undertaken in those two domains that have seen the most study: board games, and 2D graphical-logic games.

---

[3] Some interesting future work lies in modular encodings: instead of choosing a specific genre, a generator might pick and choose a generative space consisting of a combat system, 2D grid movement, an inventory system, etc. [19].

6.3 Board games

Board games were the first domain in which systems were built to procedurally generate game rules. They have several features that make them a natural place to start. For one, there is a discrete, finite structure to the games that simplifies encoding; unlike computer games, which are defined by an often complex body of code, games like chess are defined by simple sets of rules. Secondly, there is already a culture of inventing board-game variants, so automatic invention of game variants can draw from existing investigations into *manual* generation of game variants, and the design books that have been written about those investigations [8, 2].

6.3.1 Symmetric, chess-like games

The earliest rule-generating system, predating the more recent resurgence in PCG research, was METAGAME [27], which generated "symmetric, chess-like games". The *chess-like* part means that the games take place on a grid, and are structured around two players taking turns moving pieces according to certain rules; these pieces can also be removed from the board in certain circumstances. The *symmetric* part means that the two players start on opposite ends of the board with symmetric starting configurations, and all game rules are identical for each player, just flipped to the other side of the board. For example, if METAGAME invented a chess variant in which pawns could capture sideways, this would always be true for both the black and white player; the space of games METAGAME represents doesn't include asymmetric games where players start with different pieces, or make moves according to differing rules.

The symmetric aspect of the game rules is enforced by construction: only one set of rules is encoded in the generator, and those rules are applied to both players, so any change to an encoded rule automatically changes the rules for both sides. The space of possible rules is encoded in a hierarchical *game grammar* that specifies options for the board layout, how pieces can move, how they can capture, winning conditions, and so on. Specific games are generated by simply stochastically sampling from that grammar, and then imposing some checks for basic game playability. Note that this is a constructive rather than a search-based approach; the system does not test the quality of generated games as part of the generation process, and does not search the space of games it can express as much as it randomly samples it. The generator also has a few parameter knobs available, allowing the user to tweak some aspects of what's likely to be generated, such as the average complexity of movement rules.

Pell's motivation for building METAGAME was not game generation itself, but testing AI systems on the problem of general game playing. By the early 1990s, there was a worry that computer chess competitions were causing researchers to produce systems *so* specifically engineered to play chess and only chess, that they might no longer be advancing artificial intelligence in general. Pell proposed that more fundamental advances in AI would be better served by forcing game-playing AI systems to play a wider space of games, where they wouldn't know all the rules in advance, and couldn't hard-code as many details of each specific game [26]. To actually set up such a competition, he needed a way to define a large space of games, and a generator that could produce specific games from that space, to send to the competing systems. METAGAME was created to provide that more general space of test games, and as a result, also became the first PCG system for game rules.

6.3.2 Balanced board games

While METAGAME generated a fairly wide range of games, the end result was controllable only implicitly: games were not selected for specific properties, but chosen randomly from the game grammar.

One property that is frequently desired in symmetric games is game balance: there shouldn't be a large advantage for one side or the other, such that the outcome is too strongly determined by who starts with the white pieces versus the black pieces. METAGAME produces games that are *often* balanced by virtue of having symmetric rule sets, which tend to produce balanced gameplay. But a symmetric rule set does not automatically mean a game will be balanced: moving first can often be a large advantage, or it might even in some cases be a disadvantage. Hom and Marks [13] decided to address the goal of balance directly. They first took a much smaller space of chess variants, to allow the space to be more exhaustively searched. Then, they evaluated candidate games for balance by having computer players play against each other a number of times, and rejected games with simulated win rates that deviated too far from 50/50.

This process ends up feeding the original motivating application of METAGAME back into the generation of game rules. METAGAME had been designed to test general game-playing agents, which were new at the time. Over the years, a number of research and commercial systems were developed, which could take an arbitrary game encoded in a description language, and attempt to play it. Hom and Marks took one such general game-playing system, Zillions of Games, and set it to play their generated games as a way of evaluating them.

The changes from METAGAME introduced here are fairly general ones which are seen in other PCG systems: the idea of an *evaluation function* to decide what constitutes a good example, and *simulation* as a way of specifying an evaluation function in a complex domain, where it's difficult to specify one directly. Here, simulation is done by the computer playing the game against itself, and the evaluation function is how close its win rate comes to being 50/50 from each side of the board.

6.3.3 Evolutionary game design

The obvious next step is to use this ability to simulate and evaluate general games to guide the automated search for new games. For example, the evaluation function (also known as the *fitness function*) can be used to direct the evolution of rule sets, to search for new combinations of mechanics that produce fit, interesting games. This section describes an experiment in evolutionary board game design called Ludi, which produced the first fully computer-invented games to be commercially published [3].

6.3.3.1 Representation

Games are described in the Ludi system as symbolic expressions in simple *ludemic* form (a *ludeme* is a unit of game information). For example, Tic-Tac-Toe is described as follows:

```
(game Tic-Tac-Toe
  (players White Black)
  (board (tiling square i-nbors) (shape square) (size 3 3))
  (end (All win (in-a-row 3)))
)
```

This game is played by two players, White and Black, on a square 3×3 board including diagonals (*i-nbors*), and is won by the first player to form three-in-a-row of their colour. By default, players take turns placing a piece of their colour on an empty board cell per turn.

The Ludi language is *procedural* rather than *declarative* in nature, being composed of high-level rule concepts rather than low-level machine instructions or logic operations, as per the Stanford GDL. This makes the language less general as every rule must be predefined by the programmer, but has the advantages of simplicity and clarity; most readers should be able to recognise the game described above despite having no prior knowledge of the system. Further, it allows rule sets to be described and manipulated as high-level conceptual units, much as humans conceptualise games when playing and designing them.

6.3.3.2 Evaluation

The Ludi system evaluates a rule set by playing the game against itself over a number of self-play trials. A rule set is deemed to be "fit" in this context if it produces a non-trivial and interesting contest for the players. The basic approach is similar to that used by Althöfer [1] and Hom and Marks [13], but in this case a much broader range of 57 aesthetic measurements are made, divided into

- *Intrinsic* criteria based directly on the rule set.
- *Playability* criteria based on the outcomes of the self-play trials.
- *Quality* criteria based on trends in play.

The intrinsic criteria measure the game at rest directly from its rule set. However, the true nature of a game does not emerge until the game is actually played, so it was not surprising that no intrinsic criteria ultimately proved useful when these criteria were correlated with human player rankings for a suite of test games.

It was found that the playability criteria, based on the game outcomes, provided a useful and robust estimate of the basic playability of a game. Four of these criteria proved particularly good at identifying unfit rule sets, constituting a *playability filter* that formed the first line of defense to quickly weed out games that

- result in draws more often than not,
- are too unbalanced towards either player,
- have a serious first- or second-move advantage, or
- are too short or too long on average.

Games that pass the playability filter are then subject to a number of more subtle and time-intensive quality measurements, based on the *lead histories* of the simulated games. The lead history of a game is a record of the difference between the estimated strength of the board position of the eventual winner and the eventual loser at each turn. Such quality measurements are more subtle and less reliable than the playability measurements, but offer the potential to capture a richer snapshot of the player experience.

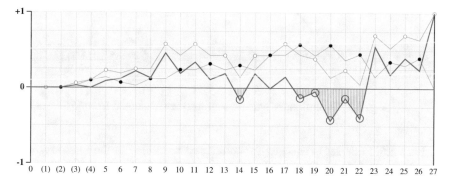

Fig. 6.1: Lead history showing drama in a game. Adapted from [5]

For example, Figure 6.1 shows the lead history of a game lasting 27 moves. The white and black dots show the players' estimated fortunes, respectively, while the red line shows the difference between them at each move. This example demonstrates a dramatic game, in which the ultimate winner (White) spends several moves in a relatively negative (losing) position before recovering to win the game. Such *drama* is a key indicator of interesting play that human designers typically strive to achieve when designing board games.

6.3.3.3 Generation

Rule sets are evolved using a *genetic programming* (GP) approach, summarised in Figure 6.2. A population of games is maintained, ordered by fitness, then for each generation a pair of relatively fit parents are selected and mated using standard *crossover* and *mutation* operations to produce a child rule set. The symbolic expressions used to describe games constitute rule trees that are ideal for this purpose.

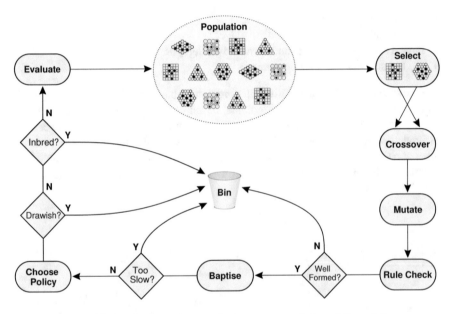

Fig. 6.2: Evolutionary game design process. Adapted from [5]

Each child rule set is checked for correctness according to the Ludi language, playability, performance and similarity to other rule sets in the population. Rule sets that pass these checks are given a unique name, officially making them a game, and are then measured for fitness and added to the population. The name for each game is also generated by the system, based on letter frequencies in a list of Tolkien-style names.

6.3.3.4 Evolved Games

Ludi evolved 1,389 new games over a week, of which 19 where deemed "playable" and two have proven to be of exceptional quality. The best of these, Yavalath, is described below:

```
(game Yavalath
   (players White Black)
   (board (tiling hex) (shape hex) (size 5))
   (end (All win (in-a-row 4)) (All lose (in-a-row 3)))
)
```

Yavalath is similar to Tic-Tac-Toe played on a hexagonal board, except that players win by making four-in-a-row (or more) of their colour but lose by making three-in-a-row beforehand. This additional condition may at first seem a redundant afterthought, but players soon discover that it allows some interesting tactical developments in play.

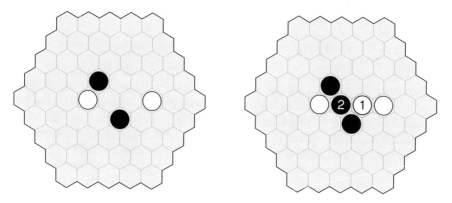

Fig. 6.3: White forces a win in Yavalath. Adapted from [4]

For example, Figure 6.3 shows a position in which White move **1** forces Black to lose with blocking move **2**. Such forcing moves allow players to dictate their opponent's moves to some extent and set up clever forced sequences. This *emergence* of complex behaviour from such simple rules provides an "aha!" moment that players find quite compelling, and is exactly what is hoped for from an evolutionary search.

The other interesting game evolved by Ludi is called Ndengrod:

```
(game Ndengrod
   (players White Black)
   (board (tiling hex) (shape trapezium) (size 7 7))
   (pieces (Piece All (moves (move
      (pre (empty to)) (action (push)) (post (capture surround))
```

```
    ) ) ) )
    (end (All win (in-a-row 5)))
  )
```

This is also an *n*-in-a-row game—this rule dominated the rule sets of evolved games—but in this case players capture enemy groups that are surrounded to have no freedom, as per Go. This rule set also demonstrates the emergence of interesting and unexpected behaviour, due to an inherent conflict between the "capture surround" and "five-in-a-row" rules, as shown in Figure 6.4.

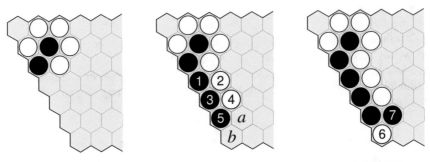

Fig. 6.4: Ladders don't work as planned in Ndengrod. Adapted from [4]

White squeezes Black against the edge to force a ladder (left), which Black must extend each turn to keep their group alive (middle). However, once the ladder reaches four pieces long after move **5**, then White cannot continue the attack at point *a* but must instead block the line at point *b*, allowing Black to escape with move **7**, and the game continues with White piece **6** now under threat.

6.3.3.5 Legacy

Yavalath and Ndengrod (renamed Pentalath) were the first fully computer-invented games to be commercially published. Yavalath was the first game released by Spanish publisher Nestorgames, and continues to be the flagship product in its catalogue of over 100 games.

Ndengrod is actually the better game of the two; it is deeper, involving a complex underlying friction between enclosure and connectivity, and is definitely more of a brain-burner. However, the more complex rules create a higher barrier to entry for beginners, hence it is destined to remain second choice. Conversely, the rules of Yavalath are intuitively obvious to any new player, and it has since been ranked in the top 100 abstract board games ever invented [4].[4]

The successful invention of board games by computer did not cause the expected backlash from players and designers. The most common response from players is

---

[4] BoardGameGeek database, August 2016 (http://www.boardgamegeek.com).

simply that they're surprised that a computer-designed game could be this simple and fun to play, while designers have so far dismissed this automated incursion into the very human art of game design as not much of a threat, as long as it produces such lightweight games. However, this attitude may change as PCG techniques—and their output—become increasingly sophisticated and challenge human experts in the field of design as well as play.

One near-miss produced by Ludi, called Lammothm, is worth mentioning to highlight a pitfall of the evolutionary approach. Lammothm is played as per Go (i.e. surround capture on a square grid) except that the aim is to connect opposite sides of the board with a chain of your pieces. Unfortunately, the evolved rule set contained the *i-nbors* attribute, meaning that pieces connect diagonally which all but ruins the game, but if this attribute is removed then the rule set suddenly becomes equivalent to that of Gonnect, one of the very best connection games [2]. Ludi was one mutation away from rediscovering a great game, but the very nature of the evolutionary process means that this mutation is not guaranteed to ever be tried for this rule set. It is possible that alternative approaches with stronger inherent local search, including *Monte Carlo tree search* (MCTS), can help address this issue.

6.3.4 Card games

Traditional card games—Poker, Uno, Blackjack, Canasta, Bridge etc.—have many features in common with board games. In particular, they are turn based and deal in discrete units (cards) which are in limited supply and can exist at any of a limited number of positions (player hands, piles etc). They also have some features that distinguish them from most board games, including not typically relying on a board and the often central importance of imperfect information (a player does not know which cards their opponents have). The limited ontology and relative ease of automated playing (due to the limited branching factor) make the domain of card games appealing for research in game generation.

Font et al. developed a description language for card games and attempted to generate card games using evolutionary search in the space defined by this language [11]. The language was defined so as to include three well-known card games—Texas Hold'em Poker, Uno and Blackjack—and implicitly games positioned between these in the game space. Initial attempts to evolve new card games in this language were made, but it was discovered that unexpectedly many of the generated games were unplayable. Efforts continue to refine the language and evaluation functions to direct the search towards playable games.

6.4 Video games

In the last few years, a small number of researchers have worked on representing and generating simple 2D graphical-logic games. By *2D graphical-logic games* we mean those games in which gameplay is based on 2D elements moving around, colliding with each other, appearing and disappearing, and the like.[5] While 2D elements moving around and colliding with each other constitutes a rather simple set of primitives out of which to build game rules, a quite large range of games can be built out of them, including such classics as *Pong*, *Pac-Man*, *Space Invaders*, *Missile Command*, and *Tetris*. These games have a different set of properties from those typically seen in board games. They are usually characterised by featuring more complex game-agent or agent-agent interaction that could easily be handled by human calculation in a board game, including semi-continuous positioning, timesteps that advance much faster than board-game turns, multiple moving NPCs, hidden state, and physics-based movement that continues even without player input. Many such games feature an avatar which the player assumes the role of and controls more or less directly, rather than selecting pieces from a board: the player "is" the Pac-Man in *Pac-Man*, which adds a new layer of interpretation [36] and experiential feeling to such games, and in turn a new axis of opportunity and challenge for rule generators.

6.4.1 *"Automatic Game Design": Pac-Man-like grid-world games*

In a 2008 paper, Togelius and Schmidhuber describe a search-based method for generating simple two-dimensional computer games [35]. The design principles of this system were that it should be able to represent a simplified discrete version of *Pac-Man*, that other games should be easy to find through simple mutations, and that the descriptions should be compact and human-readable.

The games that this system can represent all take place on a grid with dimensions 15×15 (see Figure 6.5). The grid has free space and walls, and never changes. On the grid, there is a player agent (represented in cyan in the screenshot) and *things* of three different colours (red, blue and green). Whether the things are enemies, food, helpers etc is up to the rules to define. The player agent and the things can move in discrete steps of one grid cell up, down, left or right. Each game runs for a certain number of time steps, and is won if the player reaches a score equal to or above a score threshold.

Representation: The game representation consists of a few variables and two matrices. The variables define the length of the game, the score limit, and the number of things of each colour. They also define the movement pattern of each colour. All things of a particular colour move in the same way, and the available movement patterns are standing still, moving randomly with frequent direction changes, mov-

---

[5] We borrow the term from Wardrip-Fruin [36].

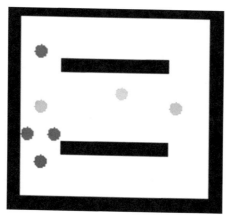

Fig. 6.5: The Automatic Game Design system by Togelius and Schmidhuber [35]

ing randomly with infrequent direction changes, moving clockwise along walls and moving counterclockwise along walls. The first of the the two matrices determines the effects of collisions between things, and between things and the agent. There is a cell for each combination of thing colours, and a cell for the combination of each colour with the player agent. The possible effects are that nothing happens, one or both things die, or one or both things teleport to a random location. For example, the matrix could specify that when a blue and a red thing collide, the blue thing dies and the red thing teleports. The other matrix is the score effects matrix. It has the same structure as the collision effects matrix, but the cells instead contain negative or positive changes to the score: for example, the player agent colliding with a blue thing might mean a score increment.

Evaluation: In the experiments described in the paper, the aim was to make games that were *learnable*. The motivation for this is the theory, introduced in various forms by psychologists such as Piaget and game designers such as Koster, that playing is learning and that a large part of the fun in games comes from learning to play them better [28, 15]. Translated to an evaluation function for game rules, the evaluation should reward games that are hard initially, but which are possible to rapidly learn to play better. Under the assumption that learnability for a machine somehow reflects learnability for a human, the evaluation function uses an evolutionary learning mechanism to learn to play games. Games that are possible to win for random players receive low fitness, whereas games that can be learnt (where the agent increases its score as much as possible) receive high fitness.[6]

---

[6] Later research has further investigated this class of fitness functions for automated game design, based on the idea that good games should be possible for intelligent agents—but not for random agents—to play well [23, 24].

6.4.2 Sculpting rule spaces: Variations Forever

All of the possible games that can be specified in a particular rule encoding make up a *generative space* of games. We've just looked at one way to explore a generative space of games and pick out interesting games from the large sea of uninteresting or even unplayable games. If we define an evaluation function to rate games from the space, we can use evolutionary computation to find games that rate highly. A different approach is to carve out interesting subsets of the space, not by rating each individual game, but by specifying properties that we want games to have, or want games to avoid. This leaves a smaller generative space with only games that satisfy the desired properties; iterative refinement can then let us zoom in on interesting areas of the generative space.

Variations Forever [31] is a game generator turned into a game, built with Answer Set Programming (ASP, see Chapter 8). In this game, the player explores different variations of game rules through playing games. The ontology and rule space is similar to but expanded compared to the rule space used in the Togelius and Schmidhuber experiment above. The games all contain things moving in a two-dimensional space, and the bulk of rules are defined by the graphical-logic effects of various types of interactions between the moving and stationary elements. However, the search mechanic is radically different. Instead of searching for rule sets that score highly on certain evaluation functions, the constraint solver finds rule sets which satisfy certain constraints. Examples of constraints include: it should be possible to win the game by pushing a red thing onto a yellow thing, or it should not be possible to lose all blue things in the game while there are still green things. These constraints are specified by the game designer, and different choices of constraints will produce larger or smaller sets of games, with different properties. The player then gets a specific game randomly chosen from that constrained space (and then another one, and then another one), and part of the game is for them to try to figure out how the rules work, and what the sequence of games have in common.

The aim of *Variations Forever* is not to produce a specific game deemed to be good, but to provide a way for game designers to define and "sculpt" generative spaces of games, where games can be included in or excluded from the space based on specific criteria. Players then explore these designer-carved generative spaces, seeing a series of games that differ in specifics but all share the specified properties.

6.4.3 Angelina

Angelina is an ongoing project by Cook and Colton to create a complete system for automatically generating novel videogames. The system has gone through several iterations, each focusing on developing a different kind of game. In the first iteration, the focus was on discrete arcade-style 2D games, and the encoding system was along the lines of the Togelius and Schmidhuber experiment above [7]. The main change is that rather than keeping the map fixed and placing the agent randomly, Angelina sees the rule set, the map, and the initial placement as three separate entities, and evolves all three of them using a form of cooperative co-evolution. This means that each different design element is evaluated partly in isolation, according to objectives which are independent of the rest of the game. However, these individual elements are also combined into full games, which are then evaluated through automated playouts to assess how well the different elements cooperate with one another. For example, a level design might show individual fitness by exhibiting a certain amount of branching or dead-ends, but be a bad fit for an object layout because it places walls over the start point for the player.

Representation: The representation of rules and mechanics in Angelina has changed through the different iterations of the software, in an attempt to increase the expressivity of the system and remove constraints on its exploration of the design space. In the first iteration of Angelina, rules were composed from a grammar-like representation of rule chunks, which produces good sets of rules, but is very dependent on the starting grammar. This is in turn dependent on the human that wrote the grammar. For Angelina, this is important because the research is partly motivated by questions of computational creativity. It's a good idea to think about issues like this when building a procedural content generator, however—if we want our systems to create things that are surprising and new, things that we could not have thought of ourselves, then it helps to consider whether our representation is constraining our systems with too many of our own preconceptions. Deciding how general or how specific your representation needs to be is a very important step in designing a generator of this kind.

To provide Angelina with more responsibility in designing the game's mechanics and rules, the second iteration of the software provided a less discrete domain for Angelina to explore. This version was focused on the design of simple Metroidvania-style platform games, where players incrementally gain powers that allow them to explore new areas of the world. Powerups are scattered through the game which change the value of one of a few hand-chosen variables in the game engine—such as the player's jump height, or the state of locked doors. The precise value associated with a given powerup was evolved as a design element in the co-evolutionary system of this version of Angelina. This meant that Angelina could make fine-grained distinctions between the player's jump height being 120 pixels or 121 pixels, which in some cases was the difference between making the player suddenly able to access the entire game world, or carefully allowing access to a small part that would provide a more natural game progression.

This notion of game mechanics as data modifiers was carried through to the next iteration of Angelina, which took the idea a step further and opened up the codebase of the underlying game engine to Angelina. This time, instead of being given a fixed set of obvious variables to choose from, Angelina was responsible for choosing both the target value *and* the target variable, out of all the variables hidden away in the entire game's code. Below is an example mechanic designed by the system. It finds the `acceleration` variable in the `player` object, and inverts the sign on its y component.

```
player.acceleration.y *= -1
```

In the Java-based game engine Flixel-GDX,[7] which Angelina uses, this is equivalent to inverting the gravitational pull on an object, similar to the gravity-flipping mechanic in Terry Cavanagh's *VVVVVV* [21]. To generate this, Angelina searched through available data fields within a sample game, and generated a type-specific modifier for it (in this case, multiplying by a negative number). This exploration of a codebase was made possible by using Java's Reflection API—a metaprogramming library that allows for the inspection, modification and execution of code at runtime. Code generation and modification is a risky business, in general—the state space can very quickly become too large to explore in any reasonable timeframe, and modifying code at runtime is similarly perilous, particularly when using something so potentially destructive as evolutionary computation.

Angelina tries to mitigate these problems in two ways: firstly, using Java as a basis for the system means that it has robust error handling. Generating and executing arbitrary code is liable to throw every kind of error imaginable. A typical run of Angelina will throw `OutOfMemoryExceptions` (by modifying data which triggers an infinite loop), `ArrayIndexExceptions` (by modifying variables which act as indexes into data structures) and `ArithmeticExceptions` (by modifying variables used in calculations, causing problems such as division by zero). However, none of these errors cause the top-level execution of Angelina to fail. Instead, they can be caught as runtime errors, and suppressed. The mechanic which caused these errors is given a low or zero fitness score, and the system then proceeds to test the next mechanic.

The second and more important way that Angelina's design overcomes issues with code generation is the evaluation criteria used to assess whether a mechanic is good or not. Figure 6.6 shows the outline for a simple level from a Mario-like platform game. The player starts the level in the red square on the left, and can run and jump. The aim is to reach the blue square on the right. We can verify that this level is unsolvable for a given jump height—the player is simply unable to scale the wall in the center of the level. This is the game configuration that Angelina begins with when evaluating a new game mechanic. The system can then add this new game mechanic to the game's codebase, and try to solve the level by reaching the exit. If Angelina is able to make progress and get to the exit, since we know the level was previously unsolvable and only the mechanic has been added we can conclude that the mechanic adds some affordance which we did not previously have. In other words, it provides some *utility* for the player.

This constraint-like evaluation approach (either the simulation reaches the exit, or it does not) is helpful in directing search through this kind of unpredictable state space. There are a few things to note about this kind of evaluation, however. Firstly, because we are generating arbitrary code modifiers, we can't give Angelina any heuristics to help it test the mechanic out. We have no idea whether a given mechanic will affect the player, enemies, the level geometry, the physics system, or whether it will outright crash the game. This means that Angelina's approach to simulating gameplay with the new mechanic is to attempt a breadth-first exhaustive

---

[7] http://www.flixel-gdx.com

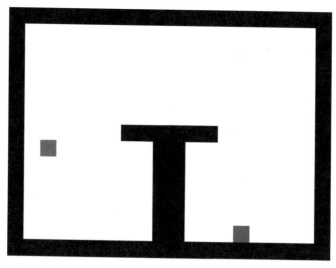

Fig. 6.6: A test level used by Angelina to evaluate generated game mechanics. The player starts in the red square on the left-hand side. They must reach the blue square on the right

simulation of gameplay. While this is almost tenable for a small example level and a restrictive move set (left, right, jump and 'special mechanic'), it's hard to imagine expanding this to a game with the complexity and scale of *Skyrim*, for example, or even *Spelunky*.

The other thing to bear in mind is that how useful a mechanic is does not necessarily relate to whether it is a good idea or not. We could imagine a very useful mechanic which automatically teleports the player to the exit, but which trivialises the rest of the game's systems entirely. Similarly, many game mechanics are specifically designed to balance utility with risk (enchanting items in *Torchlight* might result in the item being destroyed, for example) or simply exist to entertain the player. This last category is very important—mechanics such as the infinite parachute in *Just Cause 2* certainly add utility to the player's mechanical toolkit, but it is clearly designed to be enjoyable to interact with. Feelings like flow, tactility or immersion are difficult to quantify at the best of times, and are certainly not captured by the extremely utilitarian approach taken by Angelina.

Despite these shortcomings, the use of code as a domain for procedural content generation is exciting, and holds much promise. Angelina was able to rediscover many popular game mechanics, such as gravity inversion (as seen in *VVVVVV*), and bouncing (as seen in *NightSky* [22]). The purely simulation-based approach also enabled Angelina to discover obscure and nuanced emergent effects in the generated code. In one case, Angelina developed a mechanic for simple teleportation, in which the player is moved a fixed distance in a particular direction when a button is pressed. This mechanic can be used for bypassing walls, but Angelina's breadth-first simulation of gameplay also discovered that by teleporting inside a wall, it was possible to jump up out of the wall, teleport back inside, and repeat the process. This technique could be used to wall-climb—even though the game had no code relating to this feature—all made possible by a single line of code modified by Angelina.

Evaluating Angelina as an autonomous game designer has proven difficult for a number of reasons. During the development of Angelina many of the games features were static and coded by hand, such as the control schemes or (in earlier versions of the software) the game's artwork. Focusing player surveys on the aspects of games which change is difficult. Comparative testing is also difficult when the expressive range of the software is as low as it has been in some versions of Angelina. The output of the system varies within a small subgenre, which means it is difficult to make strong value judgements on whether one game is better than another, particularly mechanically. However, survey-based studies might still be the best way of getting meaningful information about the system's performance.

6.4.4 The Video Game Description Language

The Video Game Description Language (VGDL) is an effort to create a generic and flexible but compact description language for video games of the types that were seen on early home game consoles such as the Atari 2600. In this sense, it is a direct follow-up to the efforts described above (in particular Togelius and Schmidhuber), and its conceptual structure is similar. However, it is intended to be more general in that it can encode a larger range of games, and more flexible in that it decouples the description language from the game engine, the game evaluation metrics, and the generation method.

The basic design of VGDL was outlined in [10], and a first implementation of a working game engine for the language (together with several improvements to the design) was published in [30]. One of the design goals for VGDL is to be usable for general video game playing competitions, where artificial intelligence agents are tested on their capacity to play a number of games which neither the agent nor the designer of the agent has seen before [6]. These games could be manually or automatically generated, and for the idea to be viable in the long run, automatic game generation will need to be implemented at some point. The language is thus designed with ease of automatic generation in mind, though the initial stages of development have rather focused on re-implementing a range of classic games in VGDL to show the viability of doing this and test the limits of the game engine. A first iteration of the General Video Game Playing Competition[8] was run in 2014. This competition tests submitted agents against several unseen games defined in VGDL, and uses a Java-based implementation of the VGDL game engine. For future iterations, there are plans to use generated games to test agents, and to include competition tracks focused on game generation and on level generation.

---

[8] http://www.gvgai.net

A VGDL game is written in a syntax derived from Python, and is therefore relatively readable. There are four parts to a VGDL game: level mapping, sprite set, interaction set and termination set. In addition, there are level descriptions for an arbitrary positive number of levels. A level description describes a level for the game as a two-dimensional matrix of standard ASCII characters, where the level mapping defines which character maps to which type of sprite. The sprite set defines what types of sprites there are in the game and their movement behaviour, for example wall (stands still), guard robot (moves around the walls) and missile (chases the avatar). A special case is the player avatar, which the player controls directly. All sprites can obey different types of physics, such as grid-based movement or continuous movement with or without gravity. The interaction set defines most of what we call operational rules in the game, as it describes what happens when two sprites collide—similarly to the previous graphical game description efforts above, the list of possible interaction effects include death, teleportation, score increase or decrease and several others. The termination set describes various ways of ending the game, such as all sprites of a particular type disappearing, a particular sprite colliding with another etc.

6.4.5 Rulearn: Mixed-initiative game level creation

All the game generators described above have been non-interactive content generators, in that they generate a complete rule set without any human contribution. The Rulearn system by Togelius instead tries to realise interactive generation of game rules [33]. The system starts with the player controlling an agent obeying simple car physics in a 2D space containing agents of three other colours, moving randomly. Collisions will happen, but have no consequences. The player is also given an array of buttons which will effect consequences, such as "kill red", "increase score", "chase blue" and "split green". Every time the player presses a button, that consequence will happen. However, the system will also try to figure out why the player pressed that button. Using machine-learning methods on the whole history of past actions, the system will try to figure out which game the player is playing, and induce the rules behind it. The result is a mixed-initiative system for game rules, which in early testing has proved far from easy to use.

6.4.6 Strategy games

A project by Mahlmann et al. experimented with evolving key parts of strategy games [16, 17]. Strategy games are games, typically adversarial and themed on military conflict, where the player manages resources and moves units (representing e.g. tanks, soldiers and planes) around on a board. Examples include the *Civilization* series, *Advance Wars* and *Europa Universalis*; this genre of games is closely related to real-time strategy games such as *Dune II* and *StarCraft*, except for being turn based. They share characteristics with both traditional board games, such as typically being turn based and playing out on a discrete board/map, and with graphical games in the relatively complex interactions between units and in the world, more complex than could be comfortably simulated in a non-digital game.

In order to be able to generate strategy games, they developed a description language for such games, aptly called the Strategy Game Description Language (SGDL). They also developed a game engine that allows a human or a computer player to play any game described in this language. In a series of experiments, different parts of strategy games represented in this language were evolved using genetic programming. In initial experiments, the focus was on evolving how much damage each type of unit could inflict on the others in a simple strategy game with the aim of creating balanced sets of units [16]. In a later set of experiments, the complete logics for the strategy game units were evolved, with the goal of finding sets of units of balanced strengths but which were functionally different between players [17]. In these experiments several new strategy game mechanics (previously unseen to the experimenters) emerged from the experiments, including units that modified the shooting range of other units based on their proximity.

6.4.7 The future: Better languages? Better games? 3D games?

As we can see, existing work on generating graphical game has targeted games in the style of classic arcade games and home console games from the early 1980s, or simple arcade games. There is still considerable work to be done here, and nobody has yet constructed a system that can generate novel graphical games of high quality, comparable to the novel high-quality board games produced by Cameron Browne's Ludi system. One of the important open questions is how to best balance expressivity of the game description language with locality of search and density of good games; we want a representation which can represent truly novel games, but we also want that representation to be searchable. However, there are also considerable opportunities in developing game description languages that can effectively and economically describe other types of games, and game generators that take into account the specific game-design affordances and challenges that come with such games. For example, what would it take to generate playable, interesting and original FPS games? Which characteristics make games more or less easily "generatable" [34], and what techniques will best succeed at generating them, remains a wide-open research question, waiting for new experiments to push its boundaries.

6.5 Exercise: VGDL

The main theme of this chapter has been that generating rules depends heavily on how we encode rules for a particular kind of game, since these encodings define a space of games. The Video Game Description Language (VGDL) provides a fairly straightforward encoding for a set of graphical-logic games, allowing for some variation in gameplay styles, without going all the way to the intractable complexity of trying to encode every possible kind of game. In addition, it includes an interpreter and simulator in Python (pyVGDL) and another in Java (jVGDL), so that games produced in the encoding can easily be played.

This exercise is in two parts, with an open research question suggested as an optional third.

Part 1: Understanding a VGDL game. Download pyVGDL[9] or jVGDL.[10] Both packages come with a number of example games in the *examples* directory. Choose a game, and understand its encoding. You may do this by first playing it, and trying to figure out what its rules are. Then look at the rules as they're encoded in its definition file: are they the rules you figured out? Were there other rules you didn't notice? Play it again, this time with the rules in mind. Go back and forth between the written rules and the gameplay experience until you're confident you understand what happens in the game, and how that relates to what's written in the VGDL definition.

Part 2: Write a new game in VGDL. Choose a graphical-logic game suitable for representation using VGDL's vocabulary. (Many traditional arcade games of the Atari 2600 era or early Nintendo or Commodore 64 era are in their essence implementable in VGDL.) What are the objects in the game, and what rules can be written to specify the game's mechanics? You may want to start by first listing these on paper in natural language or as a set of bullet points, and then figuring out how to encode them in VGDL. You may make up your own game, or choose an existing arcade-style game to translate to VGDL.

Part 3 (optional): Write a generator that outputs VGDL games. At the time of writing, there were some preliminary experiments in doing so [23, 24], but they were not yet at the level of producing good games for humans to play. Most methods have not yet been tried; how would you approach the problem?

6.6 Summary

In order to generate game rules, you first need to devise a good representation for these rules. Several description languages for game rules have been invented. There is usually some tradeoff between the expressivity of the encoding (the range of games it can represent), the density of allowable or feasible game rules, compactness

---

[9] https://github.com/schaul/py-vgdl

[10] https://github.com/EssexUniversityMCTS/gvgai

and human readability. Once you have an encoding, you will want to find good game rules. Search-based methods have been applied in various different forms, for example by the board-game-generating Ludi system and the arcade-game-generating Angelina system. A key problem here is evaluating the quality of the rules, which is generally much harder than evaluating the quality of other kinds of game content, as the evaluation function needs to go beyond mere correctness and into some quantification of the game-theoretic and aesthetic qualities of the game. Constraint satisfaction approaches have also been used to search for good rule sets; for example Answer Set Programming was used in *Variations Forever*.

References

1. Althöfer, I.: Computer-aided game inventing. Tech. rep., Friedrich-Schiller University, Faculty of Mathematics and Computer Science (2003)
2. Browne, C.: Connection Games: Variations on a Theme. AK Peters (2005)
3. Browne, C.: Automatic generation and evaluation of recombination games. Ph.D. thesis, Queensland University of Technology (2008)
4. Browne, C.: Evolutionary Game Design. Springer, Berlin (2011)
5. Browne, C., Maire, F.: Evolutionary game design. IEEE Transactions on Computational Intelligence and AI in Games 2(1), 1–16 (2010)
6. Congdon, C., Bida, M., Ebner, M., Kendall, G., Levine, J., Lucas, S., Miikkulainen, R., Schaul, T., Thompson, T.: General video game playing. In: Dagstuhl Seminar on Artificial and Computational Intelligence in Games (2013)
7. Cook, M., Colton, S.: Multi-faceted evolution of simple arcade games. In: Proceedings of the IEEE Conference on Computational Intelligence and Games, pp. 289–296 (2011)
8. Dickins, A.: A Guide to Fairy Chess, 3rd edn. Dover (1971)
9. Dormans, J.: Simulating mechanics to study emergence in games. In: Proceedings of the 1st AIIDE Workshop on Artificial Intelligence in the Game Design Process, pp. 2–7 (2011)
10. Ebner, M., Levine, J., Lucas, S.M., Schaul, T., Thompson, T., Togelius, J.: Towards a video game description language. Dagstuhl Follow-Ups 6 (2013)
11. Font, J.M., Mahlmann, T., Manrique, D., Togelius, J.: Towards the automatic generation of card games through grammar-guided genetic programming. In: Proceedings of the 8th International Conference on the Foundations of Digital Games, pp. 360–363 (2013)
12. Genesereth, M., Love, N., Pell, B.: General game playing: Overview of the AAAI competition. AI magazine 26(2), 62 (2005)
13. Hom, V., Marks, J.: Automatic design of balanced board games. In: Proceedings of the 3rd Artificial Intelligence and Interactive Digital Entertainment Conference, pp. 25–30 (2007)
14. Khaled, R., Nelson, M.J., Barr, P.: Design metaphors for procedural content generation in games. In: Proceedings of the 2013 ACM SIGCHI Conference on Human Factors in Computing Systems, pp. 1509–1518 (2013)
15. Koster, R.: A Theory of Fun for Game Design. Paraglyph (2004)
16. Mahlmann, T., Togelius, J., Yannakakis, G.N.: Towards procedural strategy game generation: Evolving complementary unit types. Applications of Evolutionary Computation pp. 93–102 (2011)
17. Mahlmann, T., Togelius, J., Yannakakis, G.N.: Evolving card sets towards balancing Dominion. In: Proceedings of the 2012 IEEE Congress on Evolutionary Computation (2012)
18. Nelson, M.J., Mateas, M.: Towards automated game design. In: AI*IA 2007: Artificial Intelligence and Human-Oriented Computing, pp. 626–637. Lecture Notes in Computer Science 4733, Springer (2007)

19. Nelson, M.J., Mateas, M.: Recombinable game mechanics for automated design support. In: Proceedings of the Fourth Artificial Intelligence and Interactive Digital Entertainment Conference, pp. 84–89 (2008)
20. Nelson, M.J., Smith, A.M., Mateas, M.: Computational support for play testing game sketches. In: Proceedings of 5th Artificial Intelligence and Interactive Digital Entertainment Conference, pp. 167–172 (2009)
21. Nicalis: (2010). VVVVVV
22. Nicklas Nygren: (2011). NightSky
23. Nielsen, T.S., Barros, G.A.B., Togelius, J., Nelson, M.J.: General video game evaluation using relative algorithm performance profiles. In: Proceedings of the 18th Conference on Applications of Evolutionary Computation, pp. 369–380 (2015)
24. Nielsen, T.S., Barros, G.A.B., Togelius, J., Nelson, M.J.: Towards generating arcade game rules with VGDL. In: Proceedings of the 2015 IEEE Conference on Computational Intelligence and Games, pp. 185–192 (2015)
25. Orwant, J.: EGGG: Automated programming for game generation. IBM Systems Journal **39**(3–4), 782–794 (2000)
26. Pell, B.: METAGAME: A new challenge for games and learning. In: Heuristic Programming in Artificial Intelligence 3: The Third Computer Olympiad. Ellis Horwood (1992). Extended version available as University of Cambridge Computer Laboratory Technical Report UCAM-CL-TR-276.
27. Pell, B.: METAGAME in symmetric, chess-like games. In: Heuristic Programming in Artificial Intelligence 3: The Third Computer Olympiad. Ellis Horwood (1992). Extended version available as University of Cambridge Computer Laboratory Technical Report UCAM-CL-TR-277.
28. Piaget, J.: Piaget's theory. In: B. Inhelder, H.H. Chipman, C. Zwingmann (eds.) Piaget and His School: A Reader in Developmental Psychology, pp. 11–23. Springer (1976)
29. Poli, R., Langdon, W.B., McPhee, N.F.: A Field Guide to Genetic Programming (2008). http://www.gp-field-guide.org.uk
30. Schaul, T.: A video game description language for model-based or interactive learning. In: IEEE Conference on Computational Intelligence in Games (CIG), pp. 193–200 (2013)
31. Smith, A.M., Mateas, M.: Variations Forever: Flexibly generating rulesets from a sculptable design space of mini-games. In: Proceedings of the IEEE Conference on Computational Intelligence and Games, pp. 273–280 (2010)
32. Smith, A.M., Mateas, M.: Computational caricatures: Probing the game design process with AI. In: Proceedings of the 1st AIIDE Workshop on Artificial Intelligence in the Game Design Process, pp. 19–24 (2011)
33. Togelius, J.: A procedural critique of deontological reasoning. In: Proceedings of DiGRA (2011)
34. Togelius, J., Nelson, M.J., Liapis, A.: Characteristics of generatable games. In: Proceedings of the Fifth Workshop on Procedural Content Generation in Games (2014)
35. Togelius, J., Schmidhuber, J.: An experiment in automatic game design. In: Proceedings of the IEEE Symposium on Computational Intelligence and Games, pp. 111–118 (2008)
36. Treanor, M., Mateas, M., Wardrip-Fruin, N.: Kaboom! is a many-splendored thing: An interpretation and design methodology for message-driven games using graphical logics. In: Proceedings of the Fifth International Conference on the Foundations of Digital Games, pp. 224–231 (2010)

Chapter 7
Planning with applications to quests and story

Yun-Gyung Cheong, Mark O. Riedl, Byung-Chull Bae, and Mark J. Nelson

Abstract Most games include some form of narrative. Like other aspects of game content, stories can be generated. In this chapter, we discuss methods for generating stories, mostly using planning algorithms. Algorithms that search in plan space and those that search in state space can both be useful here. We also present a method for generating stories and corresponding game worlds together.

7.1 Stories in games

Games often have storylines. In some games, they are short backstories, serving to set up the action. The first-person shooter game *Doom*'s storyline, about a military science experiment that accidentally opens a portal to hell, is perhaps the canonical example of this kind of story: its main purpose is to set the mood and general theme of the game, and motivate why the player is navigating levels and shooting demons. The level progression and game mechanics have very little to do with the storyline after the game starts. In other games, the storyline structures the progression of the game more pervasively, providing a narrative arc within which the gameplay takes place. The *Final Fantasy* games are a prominent representative of this style of game storyline.

Since the theme of this book is to procedurally generate anything that goes into a game, it will not surprise the reader that we will now look at procedurally generating game storylines. As with procedural generation of game rules, discussed in the previous chapter, procedural generation of storylines is somewhat different from generation of other kinds of procedural *content*, because storylines are an unusual kind of content. They often intertwine pervasively with gameplay, and their role in a game can depend heavily on a game's genre and mechanics.

A common way of integrating a game's storyline with its gameplay, especially in adventure games and role-playing games, is the *quest* [23, 1]. In a quest, a player is given something to do in the game world, which usually is both motivated by the

© Springer International Publishing Switzerland 2016
N. Shaker et al., *Procedural Content Generation in Games*, Computational Synthesis and Creative Systems, DOI 10.1007/978-3-319-42716-4_7

current state of the storyline, and upon completion will advance it in some way. For example, the player may be tasked with retrieving an item, helping an NPC, defeating a monster, or transporting some goods to another town. Some games (especially RPGs) may be structured as one large quest, broken down into smaller sub-quests that interleave gameplay and story progression.

There are several reasons a game designer might want to procedurally generate game stories, beyond the general arguments for procedural content generation discussed in Chapter 1. One reason is that procedurally generated game worlds can lack meaning or motivation to the player, unless they are tied into the game story by procedurally generating relevant parts of the story along with the worlds. As Ashmore and Nitsche [2] argue, "without context and goals, the generated behaviours, graphics, and game spaces run the danger of becoming insubstantial and tedious." A second reason is that proceduralizing quests can make them truly *playable*. Sullivan et al. [21] note that computer RPGs often have a particularly degenerate form of quest, "generally structured as a list of tasks or milestones," rather than open-ended goals the player can creatively satisfy. Table-top RPGs have more complex and open-ended quests, since in those games, quests can be dynamically generated and adapted during gameplay by the human game-master, rather than being prewritten. Procedural quest generation gives a way to bring that flexibility back into videogame quests.

7.2 Procedural story generation via planning

One way to think about procedurally generating stories is to consider them to be a *planning* problem. In artificial intelligence, planning algorithms search for sequences of actions that satisfy a goal. A robot, for example, plans out the series of actuator movements necessary to pick up an object and carry it somewhere.

What are the sequences of actions for a story, and what is the goal? There are a number of ways to answer those questions, and researchers on procedural story generation started looking at them in the 1970s—at the time, generating purely text-based short stories, not game stories.

We could answer that a story is a sequence of events in a story world (in our case, a game world)—a sequence that eventually leads, through the chain of events, to the story's ending. Therefore we generate stories by simulating a fictional work: to tell a story, we first simulate what happens as characters move around and take actions in the story world, and then the story consists of simply recounting the events that happened. One of the first influential story-generation systems, *Tale-Spin* [14], takes this approach.

Generating stories by simulating a story world does have some shortcomings. It does not take into account what makes a *story*—particularly an interesting story—different from simply a log of events. Stories are carefully crafted by authors to have a certain pace, dramatic tension, foreshadowing, a narrative arc, etc., whereas a simulation of a day in the life of a virtual character does not necessarily have any of these features of a good story, except by accident. To solve that problem, we can look at the story-planning problem from the perspective of an author writing the story, rather than from the perspective of a protagonist taking actions in the story world. Story planning then becomes a problem of putting together a narrative sequence that fits the *author's* goals [6]. *Universe* [12] and *Minstrel* [25] are two well-known story generators that take this author-oriented approach.

For videogame stories, planning from the perspective of an author can become a more problematic concept, because players act in the game's story world, rather than in the author's head. Procedurally generating stories using an approach more like *Tale-Spin*, that takes place within the story world, can be more straightforward, since it has the advantage of talking about the same place and events that the player will be interacting with. On the other hand, we may still want a narrative arc and other author-level goals, which may lead to hybrid systems that plan author-level goals on top of story-world events [13, 19]. Many questions remain open, so procedural story generation in games is an active area of research.

In the rest of this chapter, we'll introduce the concepts and algorithms behind story planning, and walk through examples of using planning to generate interactive stories.

7.3 Planning as search through plan space

Planning can be viewed as a process that searches through a space of potential solutions to find a solution to a given problem, when knowledge about the problem domain is given. The problem is called a *planning problem* and consists of the *goal state* and the *initial state*. A solution to a planning problem is a *plan*, which contains a sequence of actions. A plan is *sound* if it reaches the goal state starting from the initial state when executed. Domain knowledge is represented as a library of *plan operators*, where each operator consists of a set of *preconditions* and a set of *effects*. Preconditions are just those conditions that must be established for the operator to be executed, and effects are just those conditions that are updated by the execution of the plan operator.

A space of potential solutions can be represented in two different ways: either as a state space or as a plan space. A *state space* can in turn be represented as a tree that consists of nodes and arcs, where a node represents a state and an arc represents a state transition by the application of an operator. The root node of the space represents the initial state when the algorithm is forward progression search while the root node represents the goal state when the algorithm is backward regression search.

Here is the pseudocode description of a state space algorithm:

```
1: construct the root node as the initial state
2: select a non-terminal node
      if non-terminal nodes are not found, return failure and exit
      if this is the goal state, return path from the
```

```
      initial to current state as solution and exit
3: select an applicable operator
   (its preconditions are true in forward progression search and
   its effects are true in backward regression search)
      if no such operators, mark node as terminal and goto 2
4: construct child nodes by applying the operator
      if the number of nodes in the graph exceeds a predefined
         maximum number of search nodes, return failure and exit
5: go to step 2
```

A *plan space* (see Figure 7.1) can be represented as a tree, which consists of nodes and arcs. Unlike a state space, however, the root node of the tree specifies the planning problem, the initial state and the goal state. Each leaf node represents a *complete plan* (i.e. solution) which can achieve the goal state from a given initial state when executed or a partial plan that cannot be refined any more due to inconsistencies in the plan. Internal nodes represent *partial plans* that contain flaws. The search process can be viewed as refining the parent node into a plan that fixes a flaw of the parent node [10]. A *flaw* in a plan can be an *open precondition* that has not been established by a prior plan step or a *threat* that can undo an established causal relationship in the plan.

Here is the pseudocode description of a partial-order planning algorithm:

```
1: construct the root node as the planning problem
2: select a non-terminal node (based on its heuristic value)
3: select a flaw in the node
      if no flaw is found, return the node as a solution and exit
4: construct children nodes by repairing the flaw
      if the flaw is an open precondition, either
         a) establish a causal link from an existing plan step, or
         b) add new plan step whose effects imply the precondition
      if the flaw is a threat, either
         a) add a temporal ordering constraint
            so that the threatened causal link is not disrupted, or
         b) add a binding constraint to separate the threatening
            step from steps involved in the threatened causal link.
      if the flaw is not repairable, mark the node as terminal
         and go to 2
      if the number of nodes in the graph exceeds a predefined
         maximum number of search nodes, return failure and exit
5: go to step 2
```

The complete plans generated by a state-space search algorithm are *total-order plans*. This means that they specify the temporal ordering constraint of every step in the plan. A *partial-order plan*, by contrast, specifies only those temporal orderings that must be established to resolve threats. For instance, imagine that you are given the goal of purchasing milk and bread in a grocery store. The goal can be successfully fulfilled without worrying about which one should be purchased first. And yet, a total-order plan specifies the order of these two purchasing actions and generates two plans: a) to purchase milk first and then purchase bread, and b) to purchase bread first and then purchase milk. On the other hand, a partial-order plan does not specify the ordering constraint and defers the decision until it is necessary.

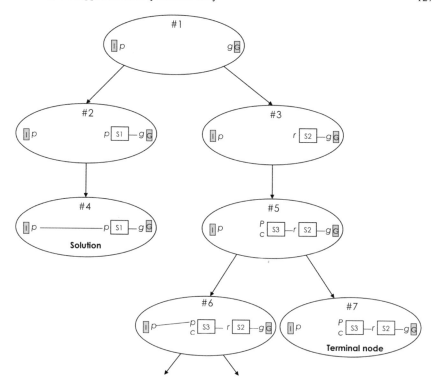

Fig. 7.1: A plan-space graph. The root node #1 represents an empty plan that contains the initial and the goal step only. The initial step contains p as an effect and the goal step contains g as its precondition. Nodes #2 and #3 are partial plans that repair the open precondition g by adding two different plan steps $S1$ and $S2$. Node #4 is a complete plan repairing an open precondition p by establishing a causal link from the initial step. The search could terminate here, if only one solution is needed. To find all solutions, the refinement search process continues from #3, generating more children (#5, #6, #7). Node #7 is marked as terminal, because there are no available operators that can repair the open precondition c. Search for additional solutions then continues from #6 (not shown)

In a plan-space search, the search process can be guided by a heuristic function which estimates the length of the optimal complete plan, based on the number of plan steps and the number of flaws that the current plan contains.

While both state-space search and plan-space search algorithms have advantages, plan-space search planners have been favoured in creating stories, because their representations are similar to the mental structure that humans construct when reading a story [24] and their search processes resemble the way humans reason to find a solution [17]. Furthermore, the causal relationships encoded in the plan structure allow further investigation of computational models of narrative, such as story summarization and affect creation [3, 5]. However, partial-order planning (POP) is computationally expensive because its space grows exponentially as the length of the plan increases. Therefore, it has not been used in many practical applications.

Hierarchical task networks (HTNs) [20, 22] represent plans hierarchically by recursively splitting composite non-primitive actions into smaller primitive actions. Figure 7.2 shows HTN *action schemas* that decompose abstract tasks into primitive tasks. HTN can be used to generate a story by generating character behaviours.

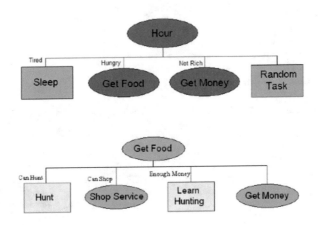

Fig. 7.2: An HTN action schema. Ovals are abstract operators, and rectangles are primitive operators. This example encodes an NPC activity that is carried out over an hour of game-world time. The NPC can sleep if tired or perform a random task. It may want to *Get Food* if hungry. *Get Food* is an abstract task is decomposed into primitive tasks such as *Hunt* and *Learn Hunting* [11]

HTN planning searches in plan-space for a suitable plan. A simple HTN algorithm is described below.

```
1: construct the root node with an abstract operator
2: select an abstract operator to expand
      if no abstract operators are found and
         all the preconditions are satisfied,
            return the network as a solution and exit
3: select an action schema whose preconditions are true
      if no such methods are found, return failure
4: decompose the abstract operator into sub-tasks
         as encoded in the action schema
5: go to step 2
```

7.4 Domain model

A *domain model* is the library of plan operator templates that encode knowledge in a particular domain (in this chapter, a story world). Various formal languages have been proposed to describe planning problems in terms of states, actions, and goals. This section focuses on two planning languages, STRIPS and ADL, which have been widely used for classical planners.

Before we get to the formalism, let us take an example. Imagine that a character in a story, named Alex, is on the rooftop of a building. His goal is to be on the ground level of the building without being injured. Alex can think of several plans immediately. For instance, Alex can take an elevator (Plan 1), can walk down the stairs (Plan 2), or can jump from the roof (Plan 3). Making the decision requires considering constraints such as his capability (e.g. Alex could be an old man having mobility problems), the building's facilities (e.g. elevators), his preference (e.g. Alex always prefers walking down the stairs for exercise), etc. If the building has an elevator and Alex wants to go to the ground level quickly, Plan 1 would be suitable. Alex may choose Plan 2 if there is no lift in the building. Alex may take Plan 3 if he has a parachute with him and a serial killer with a knife is running toward him.

The goal of planning algorithms is to formalize making these kinds of decisions: finding plans that maximise goals in the face of various conditions, constraints, and preferences. Thus, it is important to select a formal language that best expresses the problem domain.

7.4.1 STRIPS-style planning representation

STRIPS, introduced by Fikes and Nilson in 1971 [7], is the forerunner of many modern formal languages in planning. In STRIPS-style plans, a state is represented by either a *propositional literal* or a *first-order literal* where literals are ground (i.e. variable-free) and function-free. A propositional literal states a proposition which can be true or false (e.g. p, q, *PoorButler*). A first-order logic literal states a relation over objects that can be true or false (e.g. $At(Butler, House)$, $Lord(Higginbotham)$).

In STRIPS-style representations, we make a *closed-world assumption*—any conditions that are not explicitly specified are considered false. Thus only positive literals are used for the description of initial states, goal states, and preconditions. The effects of actions may include negative literals to negate particular conditions. A STRIPS-style formalization of the scenario where Alex is choosing how to exit a building (discussed above) can look like this:

- Initial state representation
  $At(Alex, Rooftop) \land Alive(Alex) \land Walkable(Rooftop, Ground) \land Person(Alex)$
  $\land Place(Rooftop) \land Place(Ground)$
- Goal State representation
  $At(Alex, Ground) \land Alive(Alex)$

- Action representation
  Action(WalkStairs (p, from, to))
  PRECONDITION: *At(p, from) ∧ Walkable(from, to) ∧ Person(p) ∧ Place(from)*
  ∧ Place(to)
  EFFECT: *¬At(p, from) ∧ At(p, to)*

In the above example, the initial state is represented by the conjunction of six first-order logic predicates. The goal state is represented by the conjunction of two predicates in the same manner. In the action representation, the action named *WalkStairs* has three variable parameters $(p, from, to)$; the action's preconditions are represented by the conjunction of five predicates; and the action's effects are denoted by the conjunction of two predicates including a negative literal. The action *WalkStairs* will be applicable and executed only when its preconditions are satisfied. After execution, the condition $At(p, from)$ will be deleted from the current state of the world and the condition $At(p, to)$ will be added to the current state of the world.

7.4.2 ADL, the Action Description Language

STRIPS is an efficient representation language for modelling states of the world. Using relatively simple logic descriptions (e.g. a conjunction of positive and function-free literals), it can convert the states and actions of a particular domain in the real world into corresponding abstract planning problems. This simplicity, however, can be a limitation in complex planning problems. Therefore many successor planning representations extend it with more features. One popular such extended language is the Action Description Language (ADL), which adds a number of additional features [16]:

- Both positive and negative literals are allowed in state descriptions, assuming open-world semantics (that is, any unspecified conditions are considered unknown, not false by default).
- Quantified variables and the combination of conjunction and disjunction are allowed in the goal state description.
- Conditional effects are allowed.
- Equality and non-equality predicates (e.g. (from ≠ to)) and typed variables (e.g. (p: Person), (from: Location)) are supported.

An ADL-style representation of the previous example is shown below:

- Initial state representation
  At(Alex, Rooftop) ∧ ¬Dead(Alex) ∧ Walkable(Rooftop, Ground) ∧ Person(Alex)
  ∧ Place(Rooftop) ∧ Place(Ground) ∧ Wearing(Alex, Parachute) ∧ ¬Injured(Alex)
  ∧ Thing(Parachute)
- Goal State representation
  At(Alex, Ground) ∧ ¬(Dead(Alex) ∨ Injured(Alex))

- Action representation
  Action(WalkStairs(p: Person, from: Place, to: Place))
  PRECONDITION: *At(p, from) ∧ (from ≠ to) ∧ (Walkable(from, to))*
  EFFECT: *¬At(p, from) ∧ At(p, to)*
  Action(JumpFromRooftop(p: Person, from: Place, to: Place, sth:Thing))
  PRECONDITION: *At(p, from) ∧ (from ≠ to) ∧ Emergent(p)*
  EFFECT: *¬At(p, from) ∧ At(p, to) ∧ (when Wearing(p, Parachute): ¬Dead(p))*

7.5 Planning a story

A story can be represented as a partial-order plan, a tuple $< S, O, C >$ where

- S is a series of events (i.e. instantiated plan operators),
- O is temporal ordering information represented as ($s1 ; s2$) where $s1$ precedes $s2$,
- C is a list of causal links where a causal link is represented by ($s, t; c$) notating a plan step s establishes c, a precondition of a step t.

Figure 7.3 illustrates a story that consists of four events that fulfills the goal $dead(Lord)$ starting from the initial state $have(Butler, Wine) ∧ have(Butler, Poison)$ $∧ serving(Butler, Lord)$. The textual description of the plan can be read as: (1) Butler puts poison in wine. (2) Butler carries wine to Lord Higginbotham. (3) Lord Higginbotham drinks wine. (4) Lord Higginbotham falls down. (The original story is from [4].)

This plan seems reasonable as a story. But is it an optimal plan that has the minimum number of steps? What if the butler gave the poison to the lord instead? Then, the plan would consist of three steps: 1) The butler carries the poison, 2) The lord drinks the poison, 3) The lord falls down.

As you may have sensed already, the new plan is logically sound but does not make a good story. Why would the lord cooperate with this plan? This is one problem that can arise with *author-centric* story generation, which may ignore individual characters' plausible intentions. An alternative approach, *character-centric* story generation, lets every character plan his/her own actions. This is more likely to produce logically consistent sets of actions, but we cannot necessarily expect that interesting stories will emerge from purely character-centric planning: A tellable situation rarely arises without the help of authorial goals. To tackle this issue, Riedl and Young proposed an intent-driven planning algorithm to balance the author-centric approach and character-centric approaches to story generation [19].

7.6 Generating game worlds and stories together

Many computer games engage players through interleaved periods of *story play* and *open-ended play*. Story play encompasses the activities of the players that promote

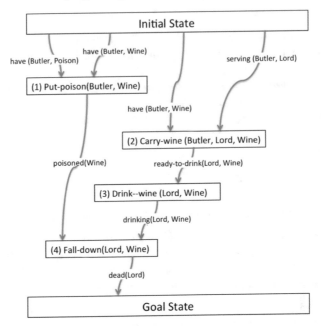

Fig. 7.3: The Butler story. A rectangle denotes an event and an arrow denotes a causal link where the event in the source establishes a condition for the event in the destination. The temporal ordering proceeds from the top to the bottom. Story originally from [4]

the progression of the game world through a narrative sequence toward a desired conclusion. As laid out in this chapter, a story can be represented as a partially ordered plan of actions that, when executed, transform the world progressively closer to a desired conclusion, represented by the goal situation. Open-ended play encompasses player activities that do not progress (nor inhibit) the story plan. Examples of open-ended play activities include exploring the spatial environment, encountering random enemies, and finding treasure or items.

This section concerns itself with the generation of playable game experiences including both story play and open-ended play. Players expect to be immersed in a *game world*, a spatial environment encompassing all locations relevant to story play and open-ended play, and inhabited by the player character and all other non-player characters. Both story play and open-ended play are often tied to the spatial environment. Unfortunately, the use of a story plan generator does not necessarily result in a playable experience without being tied to a spatial environment. In the case that a game world does not exist that suits the purposes of an automatically generated story plan, the game world may be automatically generated.

To motivate the need for game world generation, consider the fully ordered plan in Table 7.1. The plan involves a player character, the Paladin, performing a series of tasks to gain the King's trust, learn about a treasure cave, and escape a trap. Each action in the plan establishes a number of world conditions necessary for subsequent

Table 7.1: Example plan with event locations

1.	*Take* (paladin, water-bucket, palace)
2.	*Kill* (paladin, baba-yaga, water-bucket, graveyard1)
3.	*Drop* (baba-yaga, ruby-slippers, graveyard1)
4.	*Take* (paladin, shoes, graveyard1)
5.	*Gain-Trust* (paladin, king-alfred, shoes, palace)
6.	*Tell-About* (king-alfred, treasure, treasure-cave, paladin)
7.	*Take* (paladin, treasure, treasure-cave)
8.	*Trap-Closes* (paladin, treasure-cave)
9.	*Solve-Puzzle* (paladin, treasure-cave)
10.	*Trap-Opens* (paladin, treasure-cave)

actions to occur. For example, the Witch will drop her shoes only once dead, and the King will trust the Paladin once he is presented with the shoes of the Witch. A story plan only provides the essential steps to progress toward a goal situation, but does not reason about player activities that do not otherwise impact the progression of the story.

The domain model abstracts away much of the moment-to-moment activity of the player and NPCs in order to focus on the aspects of the world that are most crucial for story progression. Game play, however, is not always a sequence of discrete operations. For example, solving a puzzle may require many levers to be triggered in the right sequence. For the purposes of this chapter, we will refer to operations in a story plan as *events* to highlight their abstract nature. Events are *temporally extended*; each event can take a continuous duration of time, and there may be large durations of time between events. The plan also does not account for opportunities for open-ended play between events. For example, where is the graveyard relative to the castle, how long does it take to travel that distance, and what might the player see or experience along the way that is not directly relevant to the story plan?

If the game world is a given—i.e. there is a fixed world with a number of locations and NPCs—then there is a mapping of story events in the plan to virtual locations in the game world. For example, the game world for Table 7.1 requires a graveyard, a castle, and a treasure cave. However, due to the nature of automatically generated story plans, it is not always feasible to have a single fixed game world that meets the requirements of a story plan: locations may be missing, there may be too many irrelevant locations, or locations may need to be rearranged to make a more coherent and sensible flow. In the next section, we describe a technique to automatically generate a playable game world based on a story plan.

7.6.1 From story to space: Game world generation

Recalling that games often interleave plot points and open-ended game play, the game world to be generated must ensure a coherent sequence of events are encountered in the world. The problem can be specified as follows: given a list of events that reference locations of known types, generate a game world that allows a linear progression through the events. To map from story to space, we will utilize a metaphor of *islands* and *bridges*. Islands are areas in the spatial environment where events occur. Bridges are areas of the world between islands where open-ended game play occurs. Bridges can branch, meaning there can be areas that the player does not necessarily need to visit in the course of the story. The length of bridges and the branching factor of bridges are parameters that can be set by the designer or dictated by a player model. A game world is generated in a three-stage pipeline in which (1) a story plan is parsed for location information referenced by events, (2) an intermediate, abstract representation of the navigable space is generated, and (3) the graphical visualization of the navigable space is realized.

Table 7.2: A portion of the initial state declaration for a planning domain

Hero (paladin)	Thing (water-bucket)	Type (palace, castle)
NPC (baba-yaga)	Thing (treasure)	Type (graveyard1, graveyard)
NPC (king-alfred)	Thing (ruby-slippers)	Type (treasure-cave, cave)
Place (palace)	Evil (baba-yaga)	Type (water-bucket, bucket)
Place (graveyard1)	Type (baba-yaga, witch)	Type (ruby-slippers, shoes)
Place (treasure-cave)	Type (king-alfred, king)	Type (treasure, gold)

First, the generated story plan is parsed to extract a sequence of locations, each of which becomes an island. The story plan must be fully ordered to generate such a sequence (any partially ordered plan can be converted into a fully ordered plan). Each event in the story plan must be associated with a location. For example, in the story plan in Table 7.1, events occur at places referenced by the symbols *palace*, *graveyard1*, and *treasure-cave*. Each referenced location must have a type. This information is often found in the initial state declaration of the planning domain. Table 7.2 shows a portion of the initial state for the domain used to generate the example story plan. Thus the example story plan plays out in three locations: a castle (events 1, 5, and 6), a graveyard (events 2 through 4), and a cave (events 7 through 10).

The next stage is to generate an intermediate representation of the game world as a graph of location types called a *space tree*. A space tree is a discrete data structure that indicates how big the game world will be, how many unique locations there are, and which locations are adjacent to each other. Figure 7.4 shows an example of a space tree in which the nodes corresponding to island locations—where story plan events are to occur—are highlighted in bold and the rest of the nodes comprise the bridges.

The planning domain does not provide enough information to tell us what types of locations should be used for the bridges. We require an addition knowledge structure, called an *environment transition graph*. An environment transition graph is a data structure that captures the game designer's beliefs about good environment type transitions. Each node in an environment transition graph is a possible location type

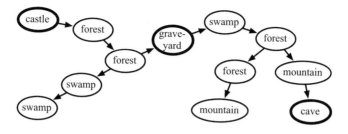

Fig. 7.4: An example space tree. Islands are marked with bold lines. Adapted from [8]

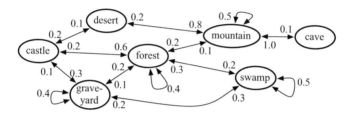

Fig. 7.5: An environment transition graph. Adapted from [8]

and edges indicate non-zero probability of transitioning from one location type to another. Figure 7.5 shows an example of an environment transition graph.

Space-tree generation can utilize any optimisation algorithm to find a space tree that meets the evaluation criteria. See Chapter 2 for the general search-based approach to procedural content generation, and [8] for specific implementation details. The evaluation criteria are:

- Whether bridges (nodes in the space tree between islands) have the preferred length.
- Whether bridges have the preferred branching factor.
- Whether the length of side paths—branch nodes that are not directly between two islands—matches the preferred side-path length.
- How closely environment type transitions between adjacent nodes match the environment transition graph probabilities.

These evaluation criteria make use of parameters set by the designer. Other evaluation criteria may be used as well.

Once the space tree has been generated via a search-based optimisation process, the third stage is to *realize* the game world graphically. The space tree gives us an abstract representation of this game world but doesn't tell us what each location should look like. Where should art assets be placed spatially to create the appearance of a forest, town, or graveyard, etc?

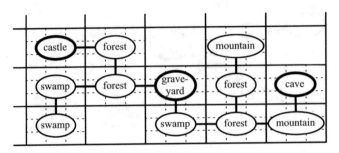

Fig. 7.6: A space tree mapped to a grid. Adapted from [8]

We describe a graphical realization process that creates a 2D, top-down, tile-based, graphical visualization of a game world described by a space tree. Starting with a grid of empty tiles, we will first map the space tree to the 2D grid and then choose tiles for each cell in the grid. If the grid is $m_{world} \times n_{world}$ tiles, then each $m_{screen} \times n_{screen}$ tiles is the number of tiles that can be displayed on the screen at any one time. Each node in the space tree will be mapped to a $m_{location} \times n_{location}$ grid of screens. In Figure 7.6, the world is 340×160 tiles, each screen is 34×16 tiles, and each location encompasses a 3×3 grid of screens (only a portion is shown). The mapping of space tree to grid is as follows. Use a depth-first traversal of the space tree, placing each child adjacent to its parent on the grid. In order to prevent an algorithmic bias toward growing the world in a certain direction (e.g. from left to right), one can randomize the order of cardinal directions in which it attempts to place each child. To minimise the likelihood that nodes will be mapped to the same portion of the grid, one can constrain the space tree such that nodes have no more than two children, for a total of three adjacent nodes. Backtrack if necessary. If there is no mapping solution, discard the space tree and resume search for the next best space tree.

Once each node in the space tree has been assigned a region on the grid, the module begins graphical instantiation of the world. Each node from the space tree has an environment type, which determines what *decorations* will be placed. Decorations are graphical assets that overlay tiles and visually depict the environment type. For a 2D tile-based realization of a game world, decorations are sprites that depict scenery found in different environment types. A forest environment has decorations consisting of grass, trees, and bushes, while a town has decorations that look like buildings, castle walls, and street paving stones.

But how does the system know where to place each decoration? This knowledge is also not present in the domain model, and a third type of external knowledge is necessary. Each environment type is associated with a function that maps decorations to a probability distribution over XY tile coordinates. We have identified two types of mapping functions.

A *Gaussian distribution* defines the dispersement of decorations around the center point of a location such that decorations are placed more densely around the

Fig. 7.7: A forest adjacent to a swamp, both with Gaussian distributions, resulting in a blended transition. Adapted from [8]

center point of each location. The advantage of a Gaussian distribution is that decorations can be placed in adjacent locations, creating the appearance that one location blends into the next, as in Figure 7.7.

A *custom distribution* is an arbitrary, designer-specified function that returns the probability of placing a decoration at any XY coordinate. Figure 7.8 shows the custom distribution for a town location type such that buildings are likely arranged in grid-like city blocks, paving stones make up streets between city blocks, and guard towers are arranged in a ring around the town perimeter.

Figure 7.9 shows an example of a complete game world with three islands extracted from Table 7.1.

7.6.2 From story to time: Story plan execution

Once the space in which the story will unfold has been generated, there are two additional issues that must be addressed: (a) the world must be populated with NPCs, and (b) the NPCs must act out the story, which is not known prior to execution. Population of the world by NPCs is a simple process of parsing the story plan for references to NPCs and instantiating sprites (based on NPC types) in the locations in which they are first required to participate in an event. Because of the temporal extension of events, NPCs must elaborate on events, including engaging in combat, engaging in dialogue, setting up and triggering traps (the world itself can be an NPC), etc. Because the story and world geometry are a-priori unknown, the NPCs must be flexible enough to elaborate on an event under a wide range of conditions based on what events preceded the current time point and how the world is laid out.

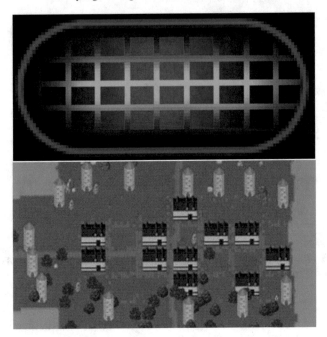

Fig. 7.8: A custom distribution for a town (above) and an example of the result (below). Brighter colour indicates greater probability of a decoration, where red indicates buildings, green indicates paving stones, and blue indicates towers. Adapted from [8]

One solution is to pair each event with a *reactive script* that decomposes the event into a number of primitive NPC behaviours. Roughly, a reactive script is an AND-OR tree structure in which internal nodes represent abstract behaviours—possibly joint between a number of characters—and leaf nodes represent primitive, executable behaviours such as animations. Reactive script execution is a walk of the tree implementing an event such that AND-nodes create sequences of sub-behaviours and OR-nodes express alternative means of decomposing achieving a behaviour, implementing *if-then-else* decision-making logic. Internal nodes may implement applicability criteria (similar to preconditions) that are used to prune sub-trees that are not supported by the state of the virtual world at execution time. Examples of reactive script technologies include behaviour trees [9], hierarchical finite state machines, hierarchical task networks [20] such as *SHOP 2* [15], and the ABL reactive behaviour planner [13].

Two types of reactive scripts are necessary to execute an automatically generated story in an open-ended game world [18]: narrative directive behaviours and local autonomous behaviours. *Narrative directive behaviours* are reactive scripts associated with event templates in the domain model. They operate as above, decomposing events into primitive behaviours. Narrative directive behaviours enact an event like

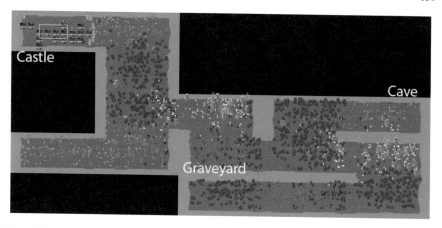

Fig. 7.9: Example game world generated from the islands in the plan in Table 7.1. Adapted from [8]

a stage manager in a play; they are not associated directly with any one character, but may control many characters at once. *Local autonomous behaviours* are associated with NPC types and execute whenever an NPC is instantiated in the world but not otherwise playing a role in an event. Local autonomous behaviours create the appearance that NPCs have rich internal lives when they are encountered by the player during open-ended play.

7.7 Lab exercise: Write a story domain model

The purpose of this exercise is to write a story domain model and characterize different planning algorithms.

1. Familiarize yourself with JSHOP2, an off-the-shelf Java implementation of the SHOP2 HTN planner (originally written in Lisp).

   - Download and install JSHOP 2.0 (http://www.cs.umd.edu/projects/shop/)
   - Check out and test the sample examples included in the package

2. Write a planning problem in terms of initial state, goal state, and actions by defining two story domains (Little Red Riding Hood and The Gift of the Magi) using either STRIPS-style or ADL-style representation. Discuss which representation is more suitable to describe the two story-world domains and explain why.
3. Convert the above planning problems into an HTN representation suitable for JSHOP2, and execute them. Discuss the strengths and weaknesses of HTN planning (or SHOP2 planner) as a story generation method/tool.
4. In the Butler story described in Section 7.5, suppose that the lord knows that the wine is poisoned and only pretends to be dead, but the butler does not know that

the lord knows. The new authorial goal is now represented as $\neg dead(Lord) \wedge arrested(Butler)$. Make a complete story plan by adding additional actions (e.g. $Call-911(Lord)$, $Arrest(Police, Butler)$), states, and causal links. Do you think that it will make the story more interesting? Why or why not?
5. Discuss the overall advantages and limitations of planning-based story generation.
6. Discuss how planning-based story-generation techniques can be effectively used in interactive storytelling systems and games.

7.8 Summary

Most games have stories, be they backstories as in a typical shooter, or stories that structure the game experience as in a role-playing game. Stories, too, can be seen as content and be generated. The most common approach to generating stories is to use some kind of planning algorithm. A planning algorithm finds a path from an initial state to a goal state; the sequence of actions that constitute this path can then be interpreted as a story. Among planning algorithms, there is a distinction between plan-space search, where the algorithm searches in the space of possible plans, and state-space search, where a plan is built up through adding new parts sequentially. A domain model is a collection of facts about the (game) world and possible actions that can be taken in it, which is then used by the planner to create a plan. There are several ways of representing a domain model, such as the STRIPS and ADL languages. For stories which have an impact on gameplay, there are ways of generating the map at the same time as the story, or the map to follow the story. Finally, search and optimisation techniques can be used to map plot points to physical locations.

References

1. Aarseth, E.: From *Hunt the Wumpus* to *EverQuest*: Introduction to quest theory. In: Proceedings of the 4th International Conference on Entertainment Computing, pp. 496–506 (2005)
2. Ashmore, C., Nitsche, M.: The quest in a generated world. In: Proceedings of the 2007 Digital Games Research Association Conference, pp. 503–509 (2007)
3. Bae, B.C., Young, R.M.: A use of flashback and foreshadowing for surprise arousal in narrative using a plan-based approach. In: Proceedings of the 1st Joint International Conference on Interactive Digital Storytelling, pp. 156–167 (2008)
4. Brewer, W., Lichtenstein, E.: Event schemas, story schemas, and story grammars. In: J. Long, A. Baddeley (eds.) Attention and Performance, vol. 9, pp. 363–379. Lawrence Erlbaum Associates (1981)
5. Cheong, Y.G., Young, R.M.: Narrative generation for suspense: Modeling and evaluation. In: First Joint International Conference on Interactive Digital Storytelling (2008)
6. Dehn, N.: Story generation after TALE-SPIN. In: Proceedings of the 7th International Joint Conference on Artificial Intelligence, pp. 16–18 (1981)
7. Fikes, R.E., Nilsson, N.J.: STRIPS: A new approach to the application of theorem proving to problem solving. Tech. Rep. 43R, SRI International (1971). SRI Project 8259

8. Hartsook, K., Zook, A., Das, S., Riedl, M.: Toward supporting storytellers with procedurally generated game worlds. In: Proceedings of the 2011 IEEE Conference on Computational Intelligence in Games, pp. 297–304. Seoul, South Korea (2011)
9. Isla, D.: Handling complexity in the Halo 2 AI. Presentation at the 2005 Game Developers Conference. URL http://www.naimadgames.com/publications/gdc05/gdc05.doc
10. Kambhampati, S., Knoblock, C.A., Yang, Q.: Planning as refinement search: A unified framework for evaluating the design tradeoffs in partial order planning. Artificial Intelligence **76**(1-2), 167–238 (1995)
11. Kelly, J.P., Botea, A., Koenig, S.: Offline planning with hierarchical task networks in video games. In: Proceedings of the 4th Artificial Intelligence and Interactive Digital Entertainment Conference, pp. 60–65 (2008)
12. Lebowitz, M.: Story-telling as planning and learning. Poetics **14**(6), 483–502 (1985)
13. Mateas, M., Stern, A.: A Behavior Language: Joint action and behavior idioms. In: H. Prendinger, M. Ishizuka (eds.) Life-like Characters: Tools, Affective Functions and Applications. Springer (2004)
14. Meehan, J.R.: The metanovel: Writing stories by computer. Ph.D. thesis, Department of Computer Science, Yale University (1976)
15. Nau, D., Ilghami, O., Kuter, U., Murdock, J.W., Wu, D., Yaman, F.: SHOP2: An HTN planning system. Journal of Artificial Intelligence Research **20**, 379–404 (2003)
16. Pednault, E.P.D.: Formulating multi-agent dynamic-world problems in the classical planning framework. In: Reasoning About Actions and Plans: Proceedings of the 1986 Workshop, pp. 47–82. Morgan Kaufmann
17. Rattermann, M.J., Spector, L., Grafman, J., Levin, H., Harward, H.: Partial and total-order planning: evidence from normal and prefrontally damaged populations. Cognitive Science **25**(6), 941–975 (2001)
18. Riedl, M.O., Stern, A., Dini, D.M., Alderman, J.M.: Dynamic experience management in virtual worlds for entertainment, education, and training. International Transactions on System Science and Applications **3**(1), 23–42 (2008)
19. Riedl, M.O., Young, R.M.: Narrative planning: balancing plot and character. Journal of Artificial Intelligence Research **39**(1), 217–268 (2010)
20. Sacerdoti, E.D.: A Structure for Plans and Behavior. Elsevier, New York (1977)
21. Sullivan, A., Mateas, M., Wardrip-Fruin, N.: Making quests playable: Choices, CRPGs, and the Grail framework. Leonardo Electronic Almanac **17**(2), 146–159 (2012)
22. Tate, A.: Generating project networks. In: Proceedings of the 1977 International Joint Conference on Artificial Intelligence, pp. 888–893 (1977)
23. Tosca, S.: The quest problem in computer games. In: Proceedings of the 1st International Conference on Technologies for Interactive Digital Storytelling and Entertainment, pp. 69–81 (2003)
24. Trabasso, T., Sperry, L.L.: Causal relatedness and importance of story events. Journal of Memory and Language **24**(5), 595 – 611 (1985)
25. Turner, S.R.: The Creative Process: A Computer Model of Storytelling and Creativity. Psychology Press (1994)

Chapter 8
ASP with applications to mazes and levels

Mark J. Nelson and Adam M. Smith

Abstract Answer set programming (ASP) is an approach to logic programming, where constraints and logical relations are declared in a Prolog-like language. ASP solvers can be used to find world configurations that satisfy constraints expressed in this language. Interestingly, many forms of content generation can be formulated as constraint-solving problems, and thus expressed in ASP. For example, maps can be represented as the position of all objects in the map, and the space of permissible maps limited by constraints expressed in the language. This chapter discusses how to use ASP for generating different types of mazes, using generation of dungeons as a running example.

8.1 What to generate and how to generate it

A common theme underlying procedural content generation is that we need to be able to specify both *what* we want our generated content to be like, and *how* to generate it. Sometimes these two parts are tightly intertwined. In the constructive methods of Chapter 3 and the fractal and noise methods of Chapter 4, we can produce different kinds of output by tweaking the algorithms until we're satisfied with their output. But if we know what properties we'd like generated content to have, it can be more convenient to directly specify what we want, and then have a general algorithm find content meeting our criteria.

The search-based framework introduced in Chapter 2 is one common way of making a content-generation algorithm general, so we can tell it what kind of content we want, and have it search for content meeting our request. An *evaluation function* specifies the properties we'd like the content to have, by numerically rating the quality of generated content according to whatever criteria we choose. A *search algorithm* then searches a space of *content encodings* to find highly rated content.

Evaluation functions summarize content quality into a single numerical rating. Then the search process, such as an evolutionary algorithm, finds content that rates

© Springer International Publishing Switzerland 2016
N. Shaker et al., *Procedural Content Generation in Games*, Computational Synthesis and Creative Systems, DOI 10.1007/978-3-319-42716-4_8

highly on that scale. Elements of an evaluation function may include both *hard constraints*—things that the content absolutely must have, such as a level being passable—and softer preferences. Evaluation of content quality may also depend on the game's mechanics. For example, whether a level is passable can depend on how a player can move, what items are available for the player's use, how enemies move, and so on; in search-based PCG this is often addressed by simulating gameplay when compute the rating.

In this chapter we look at another way of dealing with the what and how of PCG. We specify what we want our generated content to be like through *answer set programming* (ASP), a logic-programming approach. We then do the actual generation by passing the program to an ASP *solver*, which outputs content that meets the specifications of our program.

8.2 Game logic and content constraints

Instead of using a content encoding and a numerical evaluation function, here we define the *logic* of a content domain, along with *constraints* on the properties that we want the generated content to exhibit [9].

The logic of a game content domain is its structure and game mechanics. A grid-based map has a structure in which tiles are arrayed horizontally and vertically, with walls, items, structures, or other entities placed on tiles. Mechanics specify how gameplay takes place on this grid. Common mechanics include: a player starts somewhere, can move to any unoccupied adjacent square, can pick up certain kinds of items, can break certain kinds of barriers (this might require an item), etc. In short, how a game *works* makes up its logic. This logic can be encoded in computational logic [7], which means we will be able to use it to guide PCG. We don't encode how the *entire* game works, to be clear, just how the game works to the extent that it's relevant to generating the content we want.

Once we have the logic of a domain, we can write down properties that we want all generated content to have, by writing constraints that refer to the game's logic. For example: a level must have a valid path through it. What is a valid path? A sequence of moves that a player can legally make. The sequence of moves the player can legally make in turn depends on the logic of the particular game's world and rules. Some other possible constraints: all valid paths should be at least a certain minimum length, the exit and entrance must be at opposite edges of the map, and so on. We can add and remove these properties, as we think of them: perhaps the player shouldn't be able to get through a level without using at least one item (if our game has items). Maybe at least one jump should be required, or there should be a boss placed somewhere that can't be avoided. Specifying these constraints will often be done iteratively. Once we generate a few example levels, we may see things we didn't expect, and modify the set of desired properties accordingly.

The logic and constraints serve the role that the encoding and evaluation function serve in search-based PCG, but in a more explicit, symbolic form, where we've written out the logic of a game world and the properties we'd like in the generated content. The logic and constraints are then passed to a tool, called a *solver*, which solves the logic problem: it finds content that conforms to the logic of the game world and satisfies all the constraints we've specified. This approach is particularly useful when many of our desired properties are hard constraints, and may depend (perhaps in complex ways) on the game's mechanics.

8.3 Answer set programming

To apply the approach we just described in practice, we need a specific language in which to encode the game logic and constraints, and a solver for that language. In this chapter, we use answer set programming (ASP), a logic-programming approach with good support for constrained generation. While there are other possible ways to do PCG with constraint solving [6], answer set programming has a well-developed programming language with reliable existing tools, and which can be used to specify both game logic and constraints within the same language. Therefore it serves as a good general-purpose choice for programming logic- and constraint-based PCG systems.[1]

Before we jump into using ASP for a content generation task, we will first introduce some basic syntax. Answer set programs are expressed in a language called AnsProlog [1, 4], a language that visually resembles Prolog while having semantics that are more directly relatable to SAT and MAX-SAT problems.

The simplest ASP construct is a *fact*. A fact is something we declare to be true. It can be an atomic fact, which is simply a symbol that is declared true:

```
game_over.
gravity_enabled.
```

Alternatively, a fact can be specified using *predicates*, which take parameters. A predicate can be declared true for specific choices of parameters:

```
max_jump(3).
contains((2,2),wall).
```

So far, this is just a bare list of facts. We could encode a whole level this way, specifying the locations of walls, items, etc. But the interesting part comes when we add rules in addition to lists of facts. Rules specify that we can infer certain facts from others. This encodes dependencies between game elements, and also lets us start specifying dynamic elements of the game, such as game mechanics.

For example, let's say that tiles containing walls are impassable, in general. We could specify a list of facts listing explicitly which tiles are impassable. But we'd rather just say that every tile with a wall is impassable.

In conventional mathematical logic notation, we want a rule like this:

$$\forall Tile, contains(Tile, wall) \implies impassable(Tile)$$

---

[1] ASP has also been used for content generation outside of games, notably to generate music [2].

Read left to right, this says: for all tiles, if the tile contains a wall, then the tile is impassable.

In AnsProlog, this rule would be written like so:

```
impassable(Tile) :- contains(Tile,wall).
```

The symbol :- in AnsProlog is a leftward-pointing version of the implication arrow, following the convention in most programming languages that the assignment operator assigns from the right-hand side to the left. Tile is a logic variable. In AnsProlog, variables start with a capital letter, while predicates and atoms start with a lowercase letter. Variables in AnsProlog are implicitly universally quantified, so the "for-all" ($\forall$) in the mathematical version doesn't appear in the AnsProlog code.

Once we have facts and rules, that would in principle be enough to constructively generate content. However, it is typically difficult to write a set of facts and rules so that *only* content we want is derived from the rules, placing everything in exactly the right combination of places and never generating broken or undesirable output. Instead, we usually generate content in two steps. First, we constructively define a *design space*. Then we specify constraints that exclude unwanted parts of the design space.

The initial, larger design space is created by using the AnsProlog construct of *choice rules*. A choice rule specifies that the solver has an arbitrary choice in how to assign certain facts—as long as they meet some numerical constraints, and any other constraints that we might add later. The following choice rule specifies that there are between 5 and 10 walls in the level, but it doesn't specify exactly how many, or on which tiles they're located:

```
5 { contains(T,wall) : tile(T) } 10.
```

More precisely, this syntax says that, if we construct a big collection of candidate contains(T,wall) facts, for every possible T that is a tile, then the size of this set is between 5 and 10. If we have no desire to constrain the set size, we can leave off one or both of these numbers. The following choice rule simply says that a level has any number of walls:

```
{ contains(T,wall) : tile(T) }.
```

A program consisting of only the above rule produces a generative space of levels that contains any possible arrangement of walls on a grid. Of course, interesting levels require more than this. Besides adding numerical constraints on how the ASP solver makes its choices, we can exclude unwanted choices by adding different constraints that the solver must take into account. A standalone constraint is written like a rule, but has nothing on the left-hand side of the :- syntax. A solution that matches the right-hand side of the rule will be *rejected* as an invalid choice. The following example rule excludes any generated map that has a wall at $(1,1)$:

```
:- contains((1,1),wall).
```

By intermixing rules that create generative spaces, and others that prune them back down to interesting subsets, we can achieve strong control over the kind of content that is generated.

AnsProlog code is put into files with the conventional extension .lp (for "logic program"), and then passed to the solver. In this chapter we use the solver clingo from the University of Potsdam, a free and actively maintained AnsProlog solver which is part of the Potassco project of answer-set-programming tools [5].[2]

Now that we have the basic machinery of AnsProlog, we can define facts and implications, specify design spaces as free choices, and specify constraints rejecting some of those choices. We'll walk through some complete examples to show how to build and modify procedural level generators using this method.

8.4 Perfect mazes

Using our newfound ability to reason over all possible logical worlds, we will start with a simple maze generation problem. In particular, we will look at generating *perfect* mazes. A perfect maze (which may or may not actually be a desirable maze) is one in which every location is reachable while there are no closed loops. In effect, perfect mazes are trees that have been embedded into a fixed space, usually a grid.

One way to represent a tree embedded in a grid is to assign each tile in the grid a parent pointer that points to one of its adjacent cells. If the choice of parent pointers actually forms a tree, then it will be possible to traverse these pointers back to the root of the tree no matter where we start.

Let's begin by establishing a representational vocabulary for our mazes. Figure 8.1 is a self-contained AnsProlog program that uses a choice rule to assign each X/Y location a unique parent direction. This choice rule can produce facts like parent(5,7,0,-1) which might read that the tile at location $(5,7)$ has $(5,6)$ for its parent. The location $(1,1)$ will later function as the root of our tree, so we don't assign it a parent direction.

```
#const width = 5.
dim(1..width).

1 { parent(X,Y,  0,-1),
    parent(X,Y,  1,  0),
    parent(X,Y,-1,  0),
    parent(X,Y,  0,  1) } 1 :-
    dim(X), dim(Y), (X,Y) != (1,1).
```

Fig. 8.1: maze-core.lp

With just a single interesting rule, we can already begin visualizing the output of the design space we are representing so far. Using a command like the following, which uses the answer-set-solving system from the Potassco project (discussed in

---

[2] The code here is tested with clingo version 3.0.4. Most of the examples will work with minor syntax changes on clingo 4, but the --reify feature used in Section 8.6 hasn't yet been added to clingo 4, so we recommend sticking with clingo 3 when trying out the examples in this chapter.

the previous section), we can generate previews of possible mazes. Rendered examples from our program so far can be seen in Figure 8.2.

```
clingo maze-core.lp --rand-freq=1
```

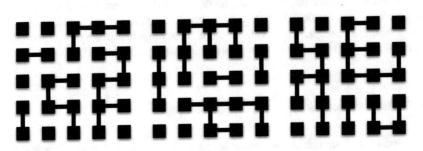

Fig. 8.2: When each tile in the maze is assigned a random parent, typical outputs show several disconnected components. Some tiles on the edges of the maze even point to a parent cell outside of the maze

To make sure we only see valid trees, we should enforce the property that the root is reachable from every tile on the grid. Figure 8.3 uses a fact, a recursive rule, and an integrity constraint to accomplish this. The linked(X,Y) property holds trivially for the root of the tree. Any tile that has a parent that is linked is linked as well. Finally, if there is some tile which does not have the linked property, something is wrong with the current assignment of parent directions and this possible world should be rejected.

```
linked(1,1).
linked(X,Y) :- parent(X,Y,DX,DY), linked(X+DX,Y+DY).

:- dim(X;Y), not linked(X,Y).
```

Fig. 8.3: maze-reach.lp

After adding these rules, we can sample examples of all and only those perfect mazes by running a command such as the following. Example outputs are shown in Figure 8.4.

```
clingo maze-core.lp maze-reach.lp
```

So far, we have used only hard constraints: tiles have exactly one parent, and every tile must be linked to the root. We can express soft constraints in AnsProlog as well by defining optimisation criteria. As an example of this for the primitive domain of mazes, let us suppose that vertical links in the maze are undesirable and that their use should be minimized. To accomplish this, the rules in Figure 8.5 define two ways of detecting a vertical link (an upward or downward parent), and the

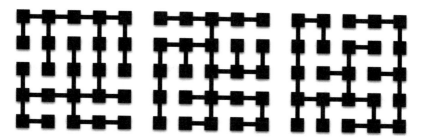

Fig. 8.4: After adding the reachability constraint for each tile, the desired tree network appears. This program captures exactly the set of all perfect mazes of a given width

`minimize` statement tells the solver that solutions which use the fewest vertical links are those that interest us. Although such statements are typically read as implying an *optimality constraint* (that only globally optimal solutions should be emitted), most answer set solvers will emit a series of answer sets they find along the way to finding one such optimal solution. By stopping the solver once it gets close enough or runs for enough time, we can implement approximate optimisation within this framework as well.

```
% soft style preferences : minimize vertical links
vertical(X,Y) :- parent(X,Y,0, 1).
vertical(X,Y) :- parent(X,Y,0,-1).
#minimize { vertical(X,Y) }.
```

Fig. 8.5: maze-bias.lp

Including the rules defining our bias against vertical links, a command like the following will allow us to sample maze designs that optimise our working evaluation criterion. Example outputs are show in Figure 8.6.

```
clingo maze-core.lp maze-reach.lp maze-bias.lp
```

8.5 Playable dungeons

Mazes are an overly simplistic example of how to carry out content generation using ASP because they can be represented with only a single kind of choice. As a slightly richer example, this section looks at generating simple dungeon maps in which a few different types of sprites are stamped down onto the familiar two-dimensional grid.

Our task will be to design a level in which the player character starts in the top-left of the grid, finds a gem in the wall of the dungeon, carries it to a central altar,

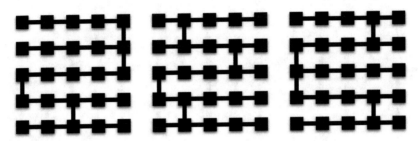

Fig. 8.6: Using the count of horizontal connections as an evaluation function, we can sample several alternative designs with a globally optimal score

where it is used to magically unlock the exit, and then walks out of that exit in the bottom right. We would like every generated level to be guaranteed to be solvable as well as to have some basic control over the pacing of the level.

To begin, examine Figure 8.7. This program establishes a vocabulary of dimension values, tiles as value pairs, and adjacency between pairs of tiles. In the character movement model we intend to capture, tiles that are one step up/down/left/right of each other are considered adjacent. A mathematical statement of this is that tile pairs with a coordinate distance of one are considered adjacent. The key part of this program is the choice rule, which states that every tile has between zero and one sprites from the set of walls, the gem, and the altar. Because we know we only want to see maps with one gem and one altar, we immediate add integrity constraints that reject those maps for which there isn't exactly one of each.

```
#const width=10.

param("width",width).

dim(1..width).

tile((X,Y)) :- dim(X), dim(Y).

adj((X1,Y1),(X2,Y2)) :-
  tile((X1,Y1)),
  tile((X2,Y2)),
  #abs(X1-X2)+#abs(Y1-Y2) == 1.

start((1,1)).
finish((width,width)).

% tiles have at most one named sprite
0 { sprite(T,wall;gem;altar) } 1 :- tile(T).

% there is exactly one altar and one gem in the whole level
:- not 1 { sprite(T,altar) } 1.
:- not 1 { sprite(T,gem) } 1.
```

Fig. 8.7: level-core.lp

Starting with these core rules, commands like the following will generate outputs like those seen in Figure 8.8.

```
clingo level-core.lp --rand-freq=1
```

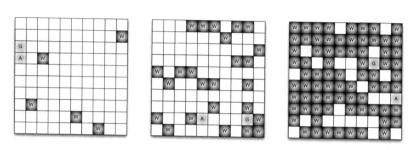

Fig. 8.8: A random result given rules that capture the basic representational vocabulary for the dungeon generation problem. A few walls (W, gray) are present, along with exactly one gem (G, green) and one altar (A, orange)

Our preliminary outputs hardly resemble interesting dungeon maps. There are many interesting maps lurking in the space we have defined, but they are hard to spot amongst the multitude of other combinations in the space. To zoom in on those maps of stylistic interest, we'll use a mixture of rules and integrity constraints to discard undesirable alteratives. A dungeon with only a sparse set of walls doesn't feel like a dungeon. A single wall sprite takes on the character of a wall when it is placed contiguously with other wall sprites. An altar should be surrounded by a few tiles of blank space, and gems should be well attached to surrounding walls. Examine Figure 8.9 for a one-line encoding of each of these concerns.

```
% style: at least half of the map has wall sprites
:- not (width*width)/2 { sprite(T,wall) }.

% style: altars have no surrounding walls for two steps
0 { sprite(T3,wall):adj(T1,T2):adj(T2,T3) } 0 :- sprite(T1,altar).

% style: altars have four adjacent tiles (not up against edge of map)
:- sprite(T1,altar), not 4 { adj(T1,T2) }.

% style: every wall has at least two neighbouring walls (no isolated rocks and spurs)
2 { sprite(T2,wall):adj(T1,T2) } :- sprite(T1,wall).

% style: gems have at least three surrounding walls (they are stuck in a larger wall)
3 { sprite(T2,wall):adj(T1,T2) } :- sprite(T1,gem).
```

Fig. 8.9: level-style.lp

With this addition, commands like the following can be used to sample stylistically valid maps such as those in Figure 8.10. Note that while the levels look reasonable locally, they are still completely undesirable on the basis that they do not support the kind of play we want—there's often not even a path from the gem to the altar, let alone from the entrance to the exit.

```
clingo level-core.lp level-style.lp
```

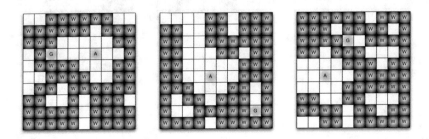

Fig. 8.10: After adding style constraints, there are many walls, the altar is surrounded by open space, and the gem is surrounded by walls on three sides. The fact that the gem is walled off is a clue that we have not yet modelled a key contraint: the level must be playable

The general strategy for ensuring we only generate playable maps is conceptually simple: generate a reference solution along with the level design. If a map contains a valid reference solution, we have a proof (by existence) that it is solvable. Even though we won't be representing the reference solution in our final output involving sprites on tiles, we can use the same language constructs as before to describe and constrain the space of possible solutions for a working map design.

Examine the rules in Figure 8.11. The key predicate is touch(Tile,State) which describes which tiles we expect the player character to touch in which gameplay state on the path to solving the level. To capture the sequence of picking up the gem, bringing it to the altar, and then exiting the level, we define three numbered states. The first rule tells us that the player will touch the start tile in state 1. From here, a series of choice rules say that touching one tile allows the player character to potentially touch any adjacent tile while retaining the same gameplay state. If the character is touching a tile containing the gem or the altar, they can transition to the next state in the sequence. The completed predicate holds (is true) if the player character touches the finish tile in the final state (after placing the gem in the altar). By rejecting every logical world where completed is not true, we zoom in on the space of different ways of solving the level. No algorithm is needed to solve a level, only a definition of what it means for a set of touched tiles to constitute a valid solution.

Although we could use the contents of Figure 8.11 as a stand-alone playability checker for human-designed dungeon maps, it is easy enough to simply use it at the same time as our previous map generator to construct a representation

```
% states :
%  1 ––> initial
%  2 ––> after picking up gem
%  3 ––> after putting gem in altar

% you start in state 1
touch(T,1) :- start(T).

% possible navigation paths
{ touch(T2,2):adj(T1,T2) } :- touch(T1,1), sprite(T1,gem).
{ touch(T2,3):adj(T1,T2) } :- touch(T1,2), sprite(T1,altar).
{ touch(T2,S):adj(T1,T2) } :- touch(T1,S).

% you can't touch a wall in any state
:- sprite(T,wall), touch(T,S).

% the finish tile must be touched in state 3
completed :- finish(T), touch(T,3).
:- not completed.
```

Fig. 8.11: level-sim.lp

of the space of maps-with-valid-solutions. A command like the following yields guaranteed-playable, styled dungeon maps like those in Figure 8.12.

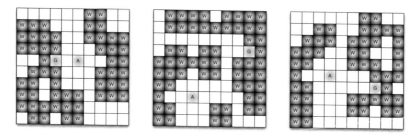

Fig. 8.12: After adding a simulation of player activity and placing constraints on the outcome, we now only see dungeon maps that have a valid solution

8.6 Constraining the entire space of play

The dungeon maps emerging from the previous section look about as good as sprite-on-grid maps containing two special objects and some walls can get. However, if we imagine playing through these maps, perhaps with simple arrow-key controls, there are still problems to resolve. In many of these maps, the task of placing the gem in the altar represents only a minor deviation from the more basic task of walking from the entrance to the exit of the dungeon. If the gem and the altar are to have any meaning for the gameplay of these maps, their placement and the arrangement of walls should conspire to make us explore the map, take detours from a start-to-finish speed run, and backtrack through familiar areas. Although each of these concerns *could* be boiled down to a set of overlapping evaluation criteria in the form of statements about the relative distances between sprites, there is a better strategy.

If our goal is to get the player to work to progress through the sequence of gameplay states, we can state a much higher-level goal. The low-level design details of the map should somehow work to make sure the player character spends at least some amount of time walking around the map in each state. How this is accomplished (with a network of rooms connected by indirect passages, perhaps) is not immediately important to us. Our high-level design goal is most directly cast as a statement about the player's experience, not the form of any particular level. We'd like to demand that, across all possible solutions to a given level design, spending a minimum amount of time in each state is unavoidable. Interpreted logically, this is a statement that is quantified over the entire space of play.

Recent advances in the use of ASP for representing design spaces now allow the direct expression of this kind of design goal. Smith et al. [8] offer a small metaprogramming library that extends normal ASP with two special predicates. Their __level_design(Atom) and __concept predicates allow the expression of a query like this: starting with a given level design and reference solution, does the design space model allow another possibility in which identical choices are made for every predicate tagged with __level_design(Atom) and in which __concept is *not* true? If so, the tagged __concept condition must not be true for the entire space of play for the given level design, and it should be rejected. The end result is a design space of level designs with reference solutions in which __concept is an *unavoidable* condition across all alternative solutions to the level. As __concept could be any quantifier-free logical formula, this language extension allows the class of extended answer set programs to express any problem in the complexity class Σ_2^P (conventionally assumed to be much larger than the class *NP*).

Returning to the dungeon map generation scenario, the rules in Figure 8.13 tag the sprite(Tile,Name) predicate as uniquely defining a level and the condition of touching at least width tiles in each of the three states as the desired unavoidable condition. A command like the following, which makes use of a special *disjunctive* answer set solver capable of solving the broader class of high-complexity problems, yields outputs like those shown in Figure 8.14.

```
clingo level-core.lp \
    level-style.lp \
    level-sim.lp \
    level-shortcuts.lp \
    --reify \
| clingo - meta{,D,O,C}.lp -1 \
| clasp
```

Before we close this section, it is instructive to ask why the following simple rule doesn't achieve the same outcome. It would seem to prune away all those solutions in which the player doesn't spend enough time in each state.

```
:- width { touch(T,1) }, width { touch(T,2) }, width { touch(T,3) }.
```

```
% holding sprites constant, ensure every solution touches at least width tiles in each state
__level_design(sprite(T,Name)) :- sprite(T,Name).
__concept :-
  width { touch(T,1) },
  width { touch(T,2) },
  width { touch(T,3) }.
```

Fig. 8.13: level-shortcuts.lp

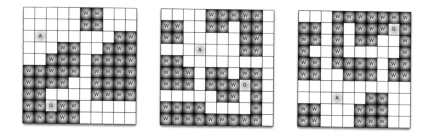

Fig. 8.14: Ensuring that the player cannot avoid spending a certain amount of time in each state has interesting emergent effects. Certain patterns that we might expect in human-crafted designs, such as the presence of hidden rooms off the main path through the level, occur naturally as the solver searches for the form of a level that gives rise to our requested function at a higher level

This integrity constraint works like the ":- not completed." rule from before. It works to make sure we only observe solutions (choices for touch(Tile,State)) that demonstrate an interesting property. Zooming in on solutions that complete the level doesn't preclude the player from choosing *not* to complete the level by simply wasting time before quitting. Likewise, zooming in on solutions in which the player wanders for a while doesn't imply that the wandering was inescapable. If we were to use this rule instead of the __level_design/__concept construction, we would most likely see many more examples like those from the previous section (Figure 8.12). In every example, it would be *possible* to wander and backtrack, but it would be unlikely to be actually required.

The idea of casting the most important properties of a level design as statements *quantified over the entire space of play* was first developed in the context of the educational puzzle game *Refraction*. What makes a given *Refraction* level desirable and relevant to its location in a larger level progression is strongly tied to which spatial and mathematical problem-solving skills *must* be exercised to solve the level, even if the level admits many possible solutions. The idea of defining a level progression primarily on the basis of which concepts are required in which levels was the basis for one of the direct-manipulation controls in the mixed-initiative progression design tool for *Refraction* [3].

Answer set programming is not the only way to write down constraints on which kinds of gameplay must be possible (e.g. a level should be solvable) and which properties of gameplay are required (e.g. that a certain skill is exercised). The key strategy to follow is to generate not just a minimal description of the content of interest, but also a description of how the content can be used towards its desired function (such as a reference solution). Many interesting properties of a piece of game content are most naturally expressed as criteria that refer to how the content is used, as opposed to any direct properties of the content itself: a good level is one that produces desired gameplay when used together with a particular game's mechanics. Despite the fact that generating content under universally quantified constraints maps to extremely high-complexity search and optimisation problems, many of these problems can be solved, in practice, in short enough times to power interactive design tools and responsive online content generators embedded into games. The use of ASP as a generation technique provides a declarative modelling language that separates the designer of a content generator from the design of the search algorithms that will be applied to these complex problems.

8.7 Exercises: Elaborations in dungeon generation

1. Run each of the examples from the text on your own machine.
2. Add a new style constraint. Make sure you understand how it changes the maps that are generated.
3. Add a new type of tile sprite, call it `lava`, that can only be traversed after the player character has touched the special `boots` tiles.
4. Change the generator so that it can be initialized with a partial map, and the generator only fills in unconstrained tiles in a way that fits style constraints.
5. Separate the playability checker from the rest of the dungeon generation program. Now apply it as a "machine playtester" [10] to point out playability flaws in levels you create yourself.
6. Design question: In the previous exercise, you took a playability checker whose initial job was to say "I wouldn't let a PCG system generate this level," and adapted it to say, "you, human designer, might have some flaws in this level you showed me." Are these really answering the same question? If you were writing a playability checker specifically to comment on human designers' levels, would you have written it differently? (See also Chapter 11.)

8.8 Summary

Answer set programming allows you to describe constraints and logical relations in a language called AnsProlog, and use an ASP solver to find all the world configurations (the "answer set") that are compatible with the expressed relations and constraints. AnsProlog is a declarative language that is syntactically similar to Prolog; however, its interpretation is different from Prolog and very different from most programming languages. ASP can be used for content generation by expressing the constraints the content must adhere to in AnsProlog, and using an ASP solver to find all configurations that adhere to these constraints. Each answer in the answer set is then treated as an individual content artifact. For example, when generating mazes, constraints may include the existence of starting points, goals, and different types of cells, as well as the existence of certain through the level. Constraints such as the reachability constraint can be implemented recursively. By building on combinations of simpler constraints and rules, complex constraints can be formulated that lead to the emergence of interesting level-design properties.

References

1. Baral, C.: Knowledge Representation, Reasoning, and Declarative Problem Solving. Cambridge University Press (2003)
2. Boenn, G., Brain, M., De Vos, M., ffitch, J.: Automatic music composition using answer set programming. Theory and Practice of Logic Programming **11**(2–3), 397–427 (2011)
3. Butler, E., Smith, A.M., Liu, Y.E., Popovic, Z.: A mixed-initiative tool for designing level progressions in games. In: Proceedings of the 26th ACM Symposium on User Interface Software and Technology, pp. 377–386 (2013)
4. Gebser, M., Kaminski, R., Kaufmann, B., Schaub, T.: Answer Set Solving in Practice. Morgan and Claypool (2012)
5. Gebser, M., Kaufmann, B., Kaminski, R., Ostrowski, M., Schaub, T., Schneider, M.: Potassco: The Potsdam answer set solving collection. AI Communications **24**(2), 107–124 (2011)
6. Horswill, I.D., Foged, L.: Fast procedural level population with playability constraints. In: Proceedings of the Eighth Artificial Intelligence and Interactive Digital Entertainment Conference, pp. 20–25 (2012)
7. Nelson, M.J., Mateas, M.: Recombinable game mechanics for automated design support. In: Proceedings of the Fourth Artificial Intelligence and Interactive Digital Entertainment Conference, pp. 84–89 (2008)
8. Smith, A.M., Butler, E., Popović, Z.: Quantifying over play: Constraining undesirable solutions in puzzle design. In: Proceedings of the Eighth International Conference on the Foundations of Digital Games, pp. 221–228 (2013)
9. Smith, A.M., Mateas, M.: Answer set programming for procedural content generation: A design space approach. IEEE Transactions on Computational Intelligence and AI in Games **3**(3), 187–200 (2011)
10. Smith, A.M., Nelson, M.J., Mateas, M.: Computational support for play testing game sketches. In: Proceedings of the Fifth Artificial Intelligence and Interactive Digital Entertainment Conference, pp. 167–172 (2009)

Chapter 9
Representations for search-based methods

Dan Ashlock, Sebastian Risi, and Julian Togelius

Abstract One of the key considerations in search-based PCG is how to represent the game content. There are several important tradeoffs here, including those between locality and expressivity. This chapter presents several more new and in some respects more advanced representations. These representations include several representations for dungeon levels, compositional pattern-producing networks for flowers and weapons, and a way of representing level generators themselves.

9.1 No generation without representation

As discussed in Chapter 2, representation is one of the two main problems in search-based PCG, and one of the two concerns when developing a search-based solution to a content generation problem. In that chapter, we also discussed the tradeoff between direct and indirect representations (the former are simpler and usually result in higher locality, whereas the latter yield smaller search spaces) and presented a few examples of how different kinds of game content can be represented. Obviously, the discussion in Chapter 2 has only scratched the surface with regard to the rather complex question of representation. This chapter will dig deeper, partly relying on the substantial volume of research that has been done on the topic of representation in evolutionary computation [2].

In the first section of this chapter, we will return to the topic of dungeons, and show how the choice of representation substantially affects the appearance of the generated dungeon. The next section discusses the generation of maps for paper-and-pen role-playing games in particular. After that, we discuss a particular kind of representation that has seen some success recently, namely Compositional Pattern-Producing Networks, or CPPNs. As we will see, this representation can be used for both flowers and weapons, and many things in between. Finally, we will discuss how we can represent not only the game content but the content generator itself,

© Springer International Publishing Switzerland 2016
N. Shaker et al., *Procedural Content Generation in Games*, Computational
Synthesis and Creative Systems, DOI 10.1007/978-3-319-42716-4_9

and search for good level generators in a search-based procedural procedural level generator generator.

9.2 Representing dungeons: A maze of choices

Dungeons or mazes (we mostly use the words interchangeably) are a topic that we have returned to several times during the book; the topic of most of Chapter 3 was dungeons, as well as the programming exercise in Chapter 2 and some of the examples in Chapter 8. The reasons for this are both the very widespread use of this type of content (including but certainly not limited to roguelike games) and the simplicity of mazes, allowing us to discuss and compare vastly different methods of generating mazes without getting lost in implementation details. It turns out that when searching for good mazes, the choice of representation matters in several different ways.

When the issue of representation arises, the goal is often enhanced performance. Enhanced performance could be improved search speed, creation of game features with desirable secondary properties that smooth ease of use, or simply fitting in with the existing computational infrastructure. In procedural content generation, there is another substantial impact of changing representation: appearance.

The pictures shown in Figure 9.1 are all level maps procedurally generated by similar evolutionary algorithms. Notice that they have very different appearances. The difference lies in the representation. All representations specify full and empty squares, but in different manners. The fitness function can be varied depending on the designer's goals and so is left deliberately vague.

Negative

The upper-left level map in Figure 9.1 starts with a matrix filled with ones. Individual loci in the gene specify where the upper left corner of a room goes and its length and height. The corridors are rooms with one dimension of length one. The red dots represent the position of a character's entrance and exit from the level. There is a potential problem with a random level having no connection from the entrance to the exit. If there is a large enough population and the representation length (number of rooms in each chromosome) is sufficient then the population contains many connected levels and selection can use these to optimise the level. This representation creates maps that look like mines.

Binary with content

The upper-right map in Figure 9.1 is created used a simple binary representation, but with *required content*. The large room with four pillars and the symmetric room with a closet opening north and south of it are the required content. They are specified

Fig. 9.1: Maps generated using four different representations. In reading order the representations are negative, binary with required content, positive, and binary with rotational symmetry. Adapted from [1]

in a configuration file. The first few loci of the chromosome specify the position of the required content elements. The remainder specify bits: 1=full 0=empty. The fitness function controls relative distances of the required content elements, and the entrances and exits. The required content represents elements the designer wants placed in an otherwise procedurally generated level. This representation generates maps that look like cave systems.

Positive

The lower-left map in Figure 9.1 uses a representation in which the loci specify walls. The starting position, direction, and maximum length of a wall are given as well as a behavioural control. The behavioural control is 0 or 1. If it is zero it stops when it hits another wall, if it is one it grows through the other wall. This representation generates maps that look like floor plans of buildings. The example shown uses eight directions—eliminating the diagonal directions yields an even more building-like appearance.

Binary with symmetry

This representation specifies directly, as full and empty, the squares of one quarter of the level with a binary gene. Each bit specifies the full/empty status of four squares in rotationally symmetric positions. There are a large number of possible symmetries that could be imposed. The imposition of symmetry yields a very different appearance.

9.2.1 Notes on usage

An important additional factor is that we need to ensure levels are connected. In the plain binary representation, if the probability of filling a square is 0.5 then it is incredibly unlikely that there is any path between entrance and exit. Similarly, if the length of walls in the positive representation is close to the diameter of the level, connected levels are unlikely. In both cases a trick called *sparse initialisation* is used. Setting the probability of a filled square to 0.2 or the maximum length of a wall to 5 makes almost all random levels connected. They are also, on average, very highly connected and so not very good. This leaves the problem of locating good levels to whatever technique the search algorithm uses to improve levels. In the examples shown, the crossover and mutation operators of the evolutionary algorithm found this to be quite easy.

The representations shown to illustrate the impact of changing representation are relatively simple. Figure 9.2 shows a more complex version of the positive representation with three types of walls. If there are two types of players, one of which can move through water and the other of which can move through fire, this representation permits the simultaneous generation of two mazes, stone-fire and stone-water, that can be optimised for particular tactical properties. In this case the stone-water maze is easier to navigate than the stone-fire maze.

9.3 Generating levels for a fantasy role-playing game

An under-explored application of procedural content generation is the automatic creation of pen-and-paper (i.e. played without a computer) fantasy role-playing (FRP) modules. Popular examples of fantasy role-playing games include *Dungeons*

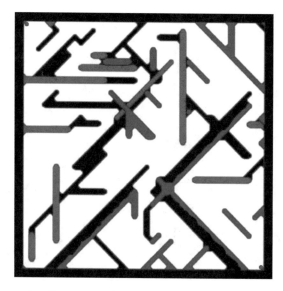

Fig. 9.2: An example of a maze, using a positive representation, with three sorts of walls: stone, fire, and water

and Dragons and the associated open gaming licence D20 systems which are used for heroic fantasy settings, *Paranoia* set in a dystopian, Orwellian future, *Champions* which is used for comic-book-style environments, and *Deadlands*, set in a haunted version of the old west. These are typically pencil-and-paper games in which players run characters and a *referee* (also known as a *game master*) interprets their actions with the help of dice, though some of these games have also been adapted into computer role-playing games. The system described here is intended to generate small adventure modules for a heroic fantasy setting.

There are a number of ways to structure generation of this type of content. The one presented here starts with required-content generation of a level. This means that the designer specifies blocks of the map, such as groups of rooms, that are forced into the level. The rest of the level is generated by filling in the area to match the relative distance between objects specified by the designer. This technique permits us to used search-based content generation to create many different levels all of which have basic properties specified by the designer. An example of a level generated in this fashion appears in Figure 9.3. Room 14 is an example of required content as is the block represented by rooms 7, 8, 11, and 12. These four rooms are a single required-content object.

Once the level map has been generated, the ACG system then automatically identifies room-sized open spaces on the map—this includes the rooms in the required content but also other spaces generated by the search algorithm optimising the level. The rooms are numbered and a combinatorial graph is abstracted from the map with rooms as vertices. The adjacency relation on the rooms is the existence of a path be-

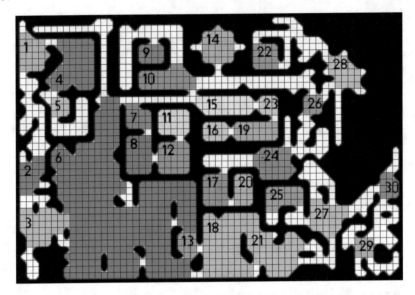

Fig. 9.3: Example of a level with automatically detected and numbered rooms

tween the rooms that does not contain a square in any other room. The graph for the map in Figure 9.3 is shown in Figure 9.4. The rooms are coloured to show which grid cells belong to them. The map with the numbered rooms, probably *sans* colours, is saved for use by the referee. The graph is handed off to the room-populating engine.

We now look at the details of the level generator. Each of these modules is an exemple and can be swapped for alternative methods with other capabilities.

9.3.1 Required content

The underlying representation for creating the levels is a simple binary one in which 1=full and 0=empty. It is modified with specifications of *required content*. An entry in the required-content configuration file looks like this:

```
12  12
111111111111
100100000001
100000000001
100111011001
100100001001
100100001001
100130001001
100111111001
100100000002
100100000002
100100000002
111111111111
```

The object specified is a 12×12 area. The representation specifies the position in the level of the upper-left-hand corner of the room, which is part of the optimisation performed by the search algorithm. The values 0,1 are mapped directly into the level, forcing values. The value 2 means that those squares are specified by the binary gene used to evolve the level. This means that some of the squares in the required content are seconded to the search algorithm. The 3 is the same as a zero—empty space— but it marks the checkpoint in the required-content object from which distances are measured. Distances are computed by dynamic programming and the fitness function uses distances between checkpoints as part of the information needed to compute fitness.

9.3.2 Map generation

The map is generated by an evolutionary algorithm. The chromosome has $2N$ integer loci for N required-content objects that are reduced modulo side length to find potentially valid places to put required-content objects. If required-content objects overlap, the chromosome is awarded a maximally bad fitness. The remainder of the position in the map, including 2's in required content, are specified by a binary gene. This gene is initialised to 20% ones, 80% zeros to make the probability the map is connected high. This is *sparse initialisation*, described earlier. The fraction of ones in population members is increased during evolution by the algorithm's crossover and mutation operators.

9.3.3 Room identification

The room identification algorithm contains an implicit definition of what a room is. The rooms appearing in the required content must satisfy this implicit definition—if not they will not be identified as rooms. For that reason a relatively simple algorithm is used to identify rooms.

Room identification algorithm

```
N=0
Scan the room in reading order
     If a 3x3 block is empty
          mark the block as in room N
          iteratively add to the room all squares with three neighbours
               already in the room
          N=N+1
     End If
End Scan
```

Once a square is marked as being part of a room, it is not longer empty, forcing rooms to have disjoint sets of squares as members. The implementation reports the squares that are members of each room and the number of squares in each room.

9.3.4 Graph generation

The rooms form the vertices of the graph of the dungeon. Earlier, a painting algorithm was used to partition space. The adjacency of rooms is computed in a very similar fashion. For each room, a painting algorithm is used to extend the room into all adjacent empty spaces until no such spaces are left. The rooms that the painting algorithm reaches are those adjacent to the room that was its focus. Each room is extended individually by painting and the paint added is erased before treating the next room. While the painting could be done simultaneously for all rooms, this might cause problems in empty spaces adjacent to more than two rooms.

The adjacency relationship has the form of a list of neighbours for each room but can be reformatted in any convenient fashion. The graph in Figure 9.4 was generated with the *GraphVis* package from an edge list—a list of all adjacent pairs of rooms.

9.3.5 Room population

The adjacency graph for the rooms is the simplest object to pass to a room population engine. The designer knows which room(s) and entrances and typically supplies this information to the population engine. The engine then does a breadth-first traversal of the graph placing lesser challenges, such as traps and smaller monsters, in the first layer, tougher monsters in the next layer, and treasure (other than that

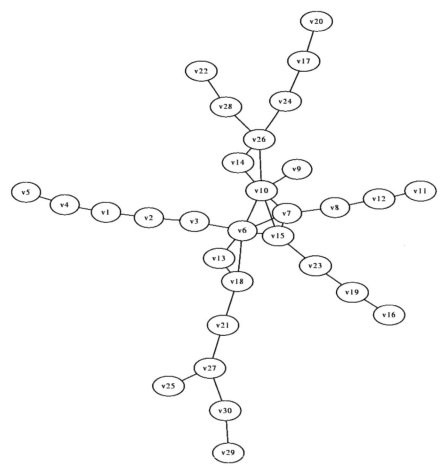

Fig. 9.4: The room adjacency graph abstraction for the level shown in Figure 9.3.
Vertex **vn** represents room **n**

carried by monsters) in the next layer. Exits to the next level are typically in the last
layer.

The required content is tagged if there is a special population engine connected
with it. An evil temple, a crypt, a dragon's den or other boss can be placed with
required content to make sure they always appear in the automatically generated
level. Correct design of the fitness function ensures that the encounters appear in an
acceptable sequence, even in a branching level, and so enable replayability.

The population engine needs a database of classified opponents, traps, and treasures scaled by difficulty. It can select randomly or in a fashion constrained by
"mood" variables. A dungeon level in a volcano, for example, might be long on fire
elementals and salamanders and short on wraiths, vampires, and other flammable

undead. A crypt, on the other hand, would be long on ghouls or skeletons and short on officious tax collectors. The creation of the encounter database, especially a careful typing system to permit enforcement of mood and style, is a critical portion of the level creation. The database needs substantially more encounters not associated with required content than it will use in a particular instance of the output of the level generator.

9.3.6 Final remarks

The FRP level generator described here is an outline. Many details can only be filled in when it is united with a particular rules system. The level generator has the potential to create multiple versions of a level and so make it more nearly replayable even when one or more of the players in a group has encountered the dungeon before. While fully automatic, the system leaves substantial scope for the designer in creating the required content and populating the encounter database.

9.4 Generating game content with compositional pattern-producing networks

In Chapter 5 we saw how grammars such as L-systems can create natural-looking plants, and learned that they are well suited to reproducing self-similar structures. In this chapter we will look at a different representation that also allows the creation of lifelike patterns, called *compositional pattern-producing networks* (CPPNs) [10]. Instead of formal grammars, CPPNs are based on artificial neural networks. In this section, we will first take a look at the standard CPPN model and then see how that representation can be successfully adapted to produce content as diverse as weapons in the game *Galactic Arms Race* [4] and flowers in the social videogame *Petalz* [7]. In *Petalz*, a special CPPN encoding enables the player to breed an unlimited number of natural-looking flowers that are symmetric, contain repeating patterns, and have distinct petals.

9.4.1 Compositional pattern-producing networks (CPPNs)

Because CPPNs are a type of artificial neural network (ANN), we will first introduce ANNs and then show that can modify them to produce a variety of different content. ANNs are computational models inspired by real brains that are capable of solving a variety of different tasks, from handwriting recognition and computer vision to robot control problems. ANNs are also applied to controlling NPCs in games and can even serve as PCG evaluation functions. For example, neural-network-based

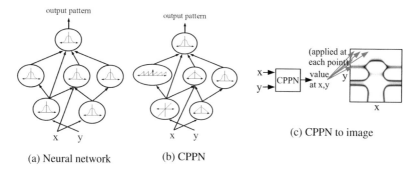

(a) Neural network (b) CPPN (c) CPPN to image

Fig. 9.5: While traditional ANNs typically only have Gaussian or sigmoid activation functions (a), CPPNs can use a variety of different function, including sigmoids, Gaussians, and sines (b). The CPPN example in this figure takes two arguments x and y as input, which can be interpreted as coordinates in two-dimensional space. Applying the CPPN to all the coordinates and drawing them with an ink intensity determined by its output results in a two-dimensional image (c). Adapted from [4]

controllers can be trained to drive like human players in a car-racing game to rate the quality of a procedurally generated track [12].

An ANN (Figure 9.5a) is an interconnected group of nodes (also called neurons) that can compute values based on external signals (e.g. infrared sensors of a robot) by feeding information through the network from its input to its output neurons. Neurons that are neither input nor output neurons are also called hidden neurons. Each neuron i has an activation level y_i that is calculated based on all its incoming signals x_j scaled by connection weights w_{ij} between them:

$$y_i = \sigma \left(\sum_{j}^{N} w_{ij} x_j \right),$$ (9.1)

where σ is called the activation function and determines the response profile of the neuron. In traditional ANNs the activation function is often the sigmoid function

$$\sigma(x) = \frac{1}{1 + e^{-kx}},$$ (9.2)

where the constant k determines the slope of the sigmoid function. The behaviour of an ANN is mainly determined by its architecture (i.e. which neurons are connected to which other neurons) and the strengths of the connection weights between the neurons.

While ANNs are usually used for control or classification problems, they can also be adapted to produce content for games. CPPNs are a variation of ANN that function similarly but can have a different set of activation functions [10]. Later we will see how special kinds of CPPNs can produce flowers in the *Petalz* videogame and weapons in GAR. While CPPNs are similar to ANNs, they have a different

Fig. 9.6: Examples of collaboratively evolved images on Picbreeder. Adapted from [9]

terminology because CPPNs are mainly used as pattern generators instead of as controllers. Let us now take a deeper look at the differences in implementation and applications between CPPNs and ANNs.

Instead of only sigmoid or Gaussian activation functions, which we can also find in ANNs (Figure 9.5a), CPPNs can include a variety of different functions (Figure 9.5b). The types of functions that we include has a strong influence on the types of patterns and regularities that the CPPN produces. Typically the set of CPPN activation functions includes a periodic function such as sine that produces segmented patterns with repetition. Another important activation function is the Gaussian, which produces symmetric patterns. Both repeating and symmetric patterns are common in nature and including them in the set of activation functions allows CPPNs to produce similar patterns artificially. Finally, linear functions can also be added to produce patterns with straight lines. The activation of a CPPN follows the ANN activation we saw in Equation 9.1, except that we now have a variety of different activation functions.

Additionally, instead of applying a CPPN to a particular input only (e.g. the position of an enemy) as is typical for ANNs, CPPNs are usually applied across a broader range of possible inputs, such as the coordinates of a two-dimensional space (Figure 9.5c). This way the CPPN can represent a complete image or as we shall see shortly also other patterns like flowers. Another advantage of CPPNs is that they can be sampled at whatever resolution is desired because they are compositions of functions. Successful CPPN-based applications include Picbreeder [9], in which users from around the Internet collaborate to evolve pictures, EndlessForms [3], which allows users to evolve three-dimensional objects, and MaestroGenesis [5], a program that enables users to generate musical accompaniments to existing songs. Figure 9.6 shows some of the images that were evolved by users in Picbreeder, which demonstrate the great variety of patterns CPPNs can represent. The CPPNs encoding these images and the other procedurally generated content in this chapter are evolved by the NEAT algorithm, which we will now examine more closely.

9.4.2 Neuroevolution of augmenting topologies (NEAT)

NEAT [11] is an algorithm to evolve neural networks; since CPPNs and ANNs are very similar, the same algorithm can also evolve CPPNs. The idea behind NEAT is that it begins with a population of simple neural networks or CPPNs that have no initial hidden nodes, and over generations new nodes and connections are added through mutations. The advantage of NEAT is that the user does not need to decide on the number of neurons and how they are connected. NEAT determines the network topology automatically and creates more and more complex networks as evolution proceeds. This is especially important for encoding content with CPPNs because it allows the content to become more elaborate and intricate over generations. While there are other methods to also evolve ANNs, NEAT is a good choice to evolve CPPNs because it worked well in the past in a variety of different domains [9, 5, 11, 4], and it is also fast enough to work in real-time environments such as interactive games.

9.4.3 CPPN-generated flowers in the Petalz videogame

Petalz [7] is a Facebook game in which procedurally generated content plays a significant role. The player can breed a collection of unique flowers and arrange them on their balconies (Figure 9.7). A flower context menu allows the player, among other things, to create new offspring through pollination of a single flower, or to combine two flower genomes together through cross-pollination. In addition to interacting with the flower evolution, the player can also post their flowers on Facebook, sell them in a virtual marketplace, or send them as gifts to other people. An important aspect of the game is that once a player purchases a flower, he can now breed new flowers from the purchased seed, and create a whole new lineage. Recently, *Petalz* was also extended with collection-game mechanics that encourage players to discover 80 unique flower species [8].

The flowers in *Petalz* are generated through a special kind of CPPN. Because the CPPN representation can generate patterns with symmetries and repetition, it is especially suited to generating natural-looking flowers with distinct petals. The basic idea behind the flower encoding is to first deform a circle to generate the shape of the flower and then to colour that resulting shape based on the CPPN-generated pattern. In contrast to the example we saw in Figure 9.5c, we now input polar coordinates $\{\theta, r\}$ into the CPPN (Figure 9.8) to generate radial flower patterns. Then we query the CPPN for each value of θ by inputting $\{\theta, 0\}$. However, instead of inputting θ into the CPPN directly, we input $sin(P\theta)$, which makes it easier for the CPPN to produce flower-like images with radial symmetry in the form of their petals. Parameter P can also be adjusted to create flowers with a different maximum numbers of petals. In the first step of the flower-generating algorithm the outline of the flower is determined, i.e. a radius value r_{max} for each θ value is calculated. In the next step, the RGB colour pattern of the flower's surface is determined by querying

Fig. 9.7: Screenshot from a *Petalz* balcony that a player has decorated with various available flower pots and player-bred flowers. Adapted from [7]

each polar coordinate between 0 and r_{max} with the same CPPN. Finally, the CPPN also allows for the creation of flowers with different layers, which reflects the fact that flowers in nature often have internal and external portions. This feature is implemented through an additional CPPN input L that determines the current layer that is being drawn. The algorithm starts by drawing the outermost layer and then each successive layer is drawn on top of the previous layers, scaled based on its depth. Because the same CPPN is determining all the layers, the different patterns can share regularities just like the different layers in real flowers.

Figure 9.9 shows examples of flowers evolved by players in *Petalz*. The CPPN-based encoding allows the discovery of a great variety of aesthetically pleasing flowers, which show varying degrees of complexity.

9.4.4 CPPN-generated weapons in Galactic Arms Race

Galactic Arms Race [4] is another successful example of a game using procedurally generated content and interactive evolution. Procedurally generated weapon projectiles, which are the main focus of this space shooter game, are evolved interactively based on gameplay data. The idea behind the interactive evolution in GAR, which was briefly discussed in Chapter 1, is that the number of times a weapon is fired is considered an indication of how much the player enjoys that particular weapon.

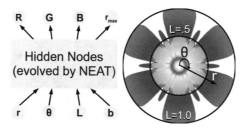

Fig. 9.8: The flower-encoding CPPN in Petalz has four inputs: polar coordinates r and θ, current layer L and bias b. The first three outputs determine the RGB colour values for that coordinate. In the first step of the algorithm the maximum radius for a given θ is determined through output r_{max}. In the next step RGB values of the flower's surface are determined by querying each polar coordinate between 0 and r_{max} with the same CPPN. The number and topology of hidden nodes is evolved by NEAT, which means that flowers can get more complex over time. From [7]

Fig. 9.9: Examples of flowers collaboratively evolved by players in the *Petalz* videogame. Adapted from [7]

As the game is played, new particle weapons are automatically generated based on player behaviour. We will now take a closer look at the underlying CPPN encoding that can generate these weapon projectiles.

Each weapon in the game is represented as a single CPPN (Figure 9.10) with four inputs and five outputs. Instead of creating a static image (Figure 9.6) or flower (Figure 9.8) the CPPNs in GAR determine the behaviour of each weapon particle over time. Each animation frame the CPPN is queried for the movement (velocity in the x and z direction) and appearance (RGB colour values) of the particle given the particle's current position in space relative to the ship (p_x, p_y) and distance d_c to its starting position. After activating the CPPN, the particles are moved to their newly determined positions and the CPPN is queried again in the next frame of animation. Evolution starts with a set of simple weapons that shoot only in a straight line and then more and more complex weapons are evolved based on the NEAT method. By adding new nodes with different activation functions, such as Gaussian and sine, interesting particle movements can evolve and the player can discover a variety of different weapons.

Figure 9.11 shows a variety of interesting weapons with vivid patterns that were evolved by players during the game. Interestingly, different weapons do not just have a different look but also tactical implications. For example, the wallmaker

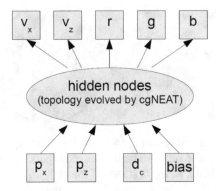

Fig. 9.10: CPPN representation of weapon projectiles in GAR. The movement of each particle is controlled by the same CPPN, which has four inputs and five outputs. The first three inputs describe the position of the particle (p_x, p_y) and the distance d_c from the location from which it was fired. After the CPPN activation, the outputs determine the particle's velocity (v_x, v_y) and RGB colour value. Adapted from [4]

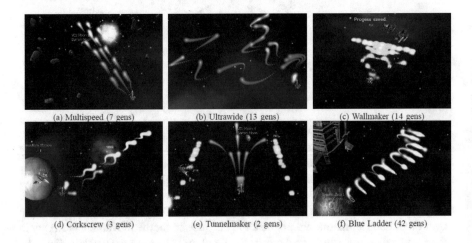

| (a) Multispeed (7 gens) | (b) Ultrawide (13 gens) | (c) Wallmaker (14 gens) |

| (d) Corkscrew (3 gens) | (e) Tunnelmaker (2 gens) | (f) Blue Ladder (42 gens) |

Fig. 9.11: Examples of CPPN-encoded weapons evolved in the *Galactic Arms Race* videogame. Adapted from [4]

weapon (Figure 9.11c) can create a wall of particles in front of the player, which allows for a more defence-oriented play. Other guns such as the multispeed weapon (Figure 9.11a) can be used in tactical situations in which a more offence-oriented approach is needed.

9.5 Generating level generators

Our final example of an advanced representation is not a representation of a particular type of game content, but rather of a level generator itself. This example, due to Kerssemakers et al. [6], views the content generator itself as a form of content, and creates a generator for it, a procedural procedural content generator generator (PPLGG). Specifically, it is a search-based generator that searches a space of generators, each of which generate levels for *Super Mario Bros.* in the Mario AI Framework.

As usual, we can understand a search-based generator in terms of representation and evaluation. The evaluation in this case is interactive: a human user looks at the various content generators, and chooses which of them (one or several) will survive and form the basis of the next generation. In order to be able to assess these content generators, the user can look at a sample of ten different levels generated by each content generator, and play any one of them; the tool also gives an estimate of how many of these levels are playable using simulation-based evaluation. Complementarily, the user can see a "cloud view" of each generator, where a number of levels generated by that generator are superimposed so that patterns shared between the levels can be seen (Figure 9.12). Figure 9.13 shows a single level in condensed view, and part of the same level in game view, where the user can actually play the level.

More interesting from the vantage point of the current chapter is the question of representation. How could you represent a content generator so as to create a searchable space of generators? In this case, the answer is that the generator is based on agents (each generator contains between 14 and 24 agents), and the generator genome consists of the parameters that define how the agents will behave. During generation, the agents move concurrently and independently, though they affect each other indirectly through the content they generate.

The genome consists of specifications for a number of agents. An agent is defined by a number of parameters, that specify how it moves, for how long, where and when it starts, how it changes the level and in response to what. The agent's behaviour is not deterministic, meaning that any collection of agents (or even any single agent) is a level generator that can produce a vast number of different levels rather than just a generative recipe for a single level.

The first five parameters below are simple numeric parameters that consist of an integer value in the range specified below. The last five parameters are categorical parameters specifying the logic of the agent, which might be associated with further parameters depending on the choice of logic. The following is a list of all parameters:

- **Spawn time [0-200]**: The step number on which this agent is put into the level. This is an interesting value as it allows the sequencing of agents, but still allows for overlap.
- **Period [1-5]**: An agent only performs movement if its lifetime in steps is divisible by the period.

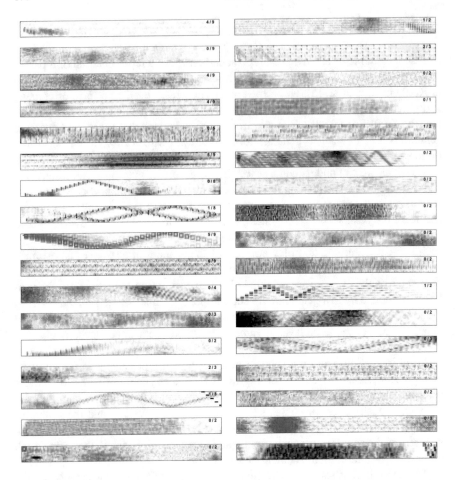

Fig. 9.12: A cloud view of several content generators. Each content generator is represented by a "cloud" consisting of multiple levels generated by that generator, overlaid on top of each other with low opacity. Adapted from [6]

- **Tokens [10-80]**: The amount of resources available to the agent. One token roughly equals a change to one tile.
- **Position [Anywhere within the level]**: The center of the spawning circle in which the agent spawns.
- **Spawn radius [0-60]**: The radius of the spawning circle in which the agent spawns.
- **Move style**: the way the agent moves every step.

    – follow a line in a specified direction (of eight possible directions) with a specified step size.
    – take a step in a random direction.

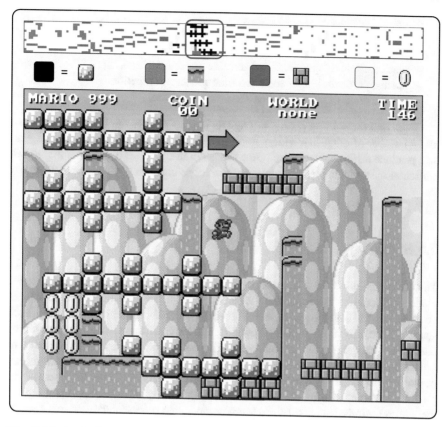

Fig. 9.13: A single generated level, and a small part of the same level in the game view. Adapted from [6]

- **Trigger state**: The condition for triggering an action, checked after each movement step.

  - always.
  - when the agent hits a specified type of terrain.
  - when a specified rectangular area is full of a specified tile type.
  - when a specified area does not contain a specified tile type.
  - with a specified probability.

- **Boundary movement**: The way the agent handles hitting a boundary.

  - bounce away.
  - go back to start position.
  - go back to within a specified rectangular area around the start position.

- **Action type**: The type of action performed if it is triggered.

- place a specified tile at position.
- place a rectangular outline of specified tiles and size around position.
- place a filled rectangle of specified tiles and size around position.
- place a circle of specified tiles and size around position.
- place a platform/line of specified tiles and size at position.
- place a cross of specified tiles and size at position.

Given that the starting position of agents implies a large amount of randomness, and a number of other behaviours imply some randomness, the same set of agents will produce different levels each time the generator is run. This is what makes this particular system a content generator generator rather than "just" a content generator.

9.6 Summary

This chapter addressed the issue of content representation within search-based PCG. How content is represented affects not only how effectively the space can be searched, but also biases the search process towards different parts of the search space. This can be illustrated by how different ways of representing a dungeon or maze yield end products that look very different, even though they are evolved to satisfy the same evaluation function and reach similar fitness. Representations can be tailored to extend the search-based paradigm in various ways, for example by providing "required content" that cannot be altered by the variation operators of the search/optimisation algorithm. More complicated representations might require a multi-step genotype-to-phenotype mapping that can be seen as a PCG algorithm in its own right. For example, compositional pattern-producing networks (CPPNs) are a form of neural network that maps position in some space to intensity, colour, direction or some other property of pixels or particles. This is an interesting content generation algorithm in itself, but can also be seen as an evolvable content representation. Taking this perspective to its extreme, we can set out to evolve actual content generators, and judge them not on any single content artefact they produce but on samples of their almost infinitely variable output. The last example in this chapter explains one way this can be done, by representing Mario AI level generators as parameters of agent-based systems and evolving those.

References

1. Ashlock, D., Lee, C., McGuinness, C.: Search-based procedural generation of maze-like levels. IEEE Transactions on Computational Intelligence and AI in Games 3(3), 260–273 (2011)
2. Ashlock, D., McGuinness, C., Ashlock, W.: Representation in evolutionary computation. In: Advances in Computational Intelligence, pp. 77–97. Springer (2012)

3. Clune, J., Lipson, H.: Evolving three-dimensional objects with a generative encoding inspired by developmental biology. In: Proceedings of the European Conference on Artificial Life (2011)

4. Hastings, E., Guha, R., Stanley, K.: Evolving content in the Galactic Arms Race video game. In: Proceedings of the IEEE Symposium on Computational Intelligence and Games, pp. 241–248 (2009)

5. Hoover, A.K., Szerlip, P.A., Norton, M.E., Brindle, T.A., Merritt, Z., Stanley, K.O.: Generating a complete multipart musical composition from a single monophonic melody with functional scaffolding. In: Proceedings of the 3rd International Conference on Computational Creativity, pp. 111–118 (2012)

6. Kerssemakers, M., Tuxen, J., Togelius, J., Yannakakis, G.N.: A procedural procedural level generator generator. In: Proceedings of the IEEE Conference on Computational Intelligence and Games, pp. 335–341 (2012)

7. Risi, S., Lehman, J., D'Ambrosio, D.B., Hall, R., Stanley, K.O.: Combining search-based procedural content generation and social gaming in the Petalz video game. In: Proceedings of the Artificial Intelligence and Interactive Digital Entertainment Conference (2012)

8. Risi, S., Lehman, J., D'Ambrosio, D.B., Stanley, K.O.: Automatically categorizing procedurally generated content for collecting games. In: Proceedings of the Workshop on Procedural Content Generation in Games (2014)

9. Secretan, J., Beato, N., D'Ambrosio, D., Rodriguez, A., Campbell, A., Folsom-Kovarik, J., Stanley, K.: Picbreeder: A case study in collaborative evolutionary exploration of design space. Evolutionary Computation 19(3), 373–403 (2011)

10. Stanley, K.O.: Compositional pattern producing networks: A novel abstraction of development. Genetic Programming and Evolvable Machines 8(2), 131–162 (2007)

11. Stanley, K.O., Miikkulainen, R.: Evolving neural networks through augmenting topologies. Evolutionary Computation 10(2), 99–127 (2002)

12. Togelius, J., De Nardi, R., Lucas, S.M.: Towards automatic personalised content creation for racing games. In: Proceedings of the IEEE Symposium on Computational Intelligence and Games, pp. 252–259 (2007)

Chapter 10
The experience-driven perspective

Noor Shaker, Julian Togelius, and Georgios N. Yannakakis

Abstract Ultimately, content is generated for the player. But so far, our algorithms have not taken specific players into account. Creating computational models of a player's behaviour, preferences, or skills is called player modelling. With a model of the player, we can create algorithms that create content specifically tailored to that player. The experience-driven perspective on procedural content generation provides a framework for content generation based on player modelling; one of the most important ways of doing this is to use a player model in the evaluation function for search-based PCG. This chapter discusses different ways of collecting and encoding data about the player, primarily player experience, and ways of modelling this data. It also gives examples of different ways in which such models can be used.

10.1 Nice to get to know you

As you play a game, you get to know it better and better. You understand how to use its core mechanics and how to combine them; you get to know the levels of the game, or, if the levels are procedurally generated, the components of the levels and typical ways in which they can be combined. You learn to predict the behaviour of other creatures, characters and systems in the game. All this you learn from your interaction from the game. While playing, you also adapt to the game: you change your behaviour so as to achieve more success in the game, or so as to entertain yourself better.

However, both you and your game take part in this interaction, and all of your interaction data is available to the game as well. In principle, the game should be able to get to know you as much as you get to know it. After all, it has seen you succeed at overtaking that other car, fail that sequence of long jumps, give up and shut down the game after crashing your plane for the seventh time or finally resort to buying extra moves after almost clearing a particular puzzle. A truly intelligent game should know how you play better than you know it yourself. And then, it

© Springer International Publishing Switzerland 2016
N. Shaker et al., *Procedural Content Generation in Games*, Computational
Synthesis and Creative Systems, DOI 10.1007/978-3-319-42716-4_10

should be able to adapt itself so as to entertain you better, or let you achieve more or less success in the game, or perhaps to give you some other kind of experience you would not otherwise have had.

The idea of *game adaptation*, the game adapting itself in response to how you play (or some other information it might have about you), is an old one. In its simplest form it is called "dynamic difficulty adjustment" (DDA), and simply means that the difficulty of the game is increased if the player does well and decreased if the player plays poorly. This can be seen in many car racing games, where the opponent cars always seem to be just ahead of you or just behind you, regardless of how well you play (also known as "rubber banding"). The game design rationale for rubber banding is that if the player is much in front of the opponents s/he will not perceive a challenge, and if the player is far behind the opponents s/he will lose hope of ever catching up; in either case, the player will likely lose interest in the game. This is sometimes rationalised as a way of keeping the player in the "flow channel". *Flow* is a concept which was invented by the psychologist Csikszentmihalyi to signify the "optimal experience", where someone is completely absorbed in the activity they are performing; one condition for this is constant but not unassailable challenge [5]. The flow concept has inspired several theories of challenge and engagement in games, such as GameFlow [27]; it is, however, limited to challenge, which is only one dimension of player experience [3].

DDA mechanisms in racing games are often implemented simply by letting the opponent cars drive faster or slower. There are interesting exceptions, such as the *Mario Kart* series, which gives more powerful power-ups to players who lag behind, some of which allow them to attack players who lead the pack. Other games might lower the difficulty of a particular section of the game after a player has failed numerous times; *Grand Theft Auto V* allows the player to simply skip any action sequence which the player has failed three times already. There are several proposals for how this could be done more automatically, using AI techniques [10]. A key realisation is that adaptation is about more than just difficulty: to begin with, difficulty is multi-dimensional, as a game can be difficult in many different ways, and people have unbalanced skill sets. The same game could be difficult for player A because of its requirement for quick reactions, for player B because of the spatial navigation, and for player C because of the nuances of the story that needs to be understood in order to solve its puzzles. Also, just having the right difficulty is in general not enough for a game to be perfectly tailored for a particular player. Different players might prefer different balances of game elements or atmospheres, such as scary, intense or contemplative parts of the game. Adaptation could in principle happen along many axes, which may not be formalised or even described. There are also many possible methods for adaptation, some of which involve modifying the content of the game or even generating new content.

In this chapter, we will focus on the use of PCG methods to adapt games to the experience of the player, which is called *experience-driven procedural content generation* [37]. Experience-driven PCG views game content as the building block of player experience which is, in turn, synthesised via content adaptation. In experience-driven PCG, a model of player experience is learned that can predict some aspect of the player's experience (e.g. challenge, frustration, engagement, spatial involvement) based on some aspect of game content. This model can then be used as a base for an evaluation function in search-based or mixed-initiative PCG. For example, a model might be learned that predicts how engaging some players think individual building puzzles are in a physics-based puzzle game. This model can then be used for evolving new puzzles, where the evaluation function rewards such puzzles that are predicted to be most engaging for the target player(s).

The chapter is structured as follows. First we describe the various ways we can elicit player experience through a game and collect information about player experience. The next section discusses algorithms for creating models of player experience, such as neuroevolutionary preference learning, based on data collected during the game interaction (model's input) and annotated player experience (model's output). A short section discusses how these models can be used in content generation, followed by a prolonged example describing experience-driven level generation in *Super Mario Bros.* in detail.

10.2 Eliciting player experience

Games can elicit rich and complex patterns of user experience as they combine unique properties such as rich interactivity and potential for multifaceted player immersion [3]. User experience in games can be elicited primarily through long- or short-term interaction with core game elements. Arguably social interaction may have a clear impact on a player's experience; however, it offers a rather challenging problem for artificial intelligence, signal processing and experience-driven PCG techniques. While an interesting direction for further research, social interaction is not included in the set of player experience elicitors considered in this chapter.

Experience-driven PCG views game content as potential *building blocks of player experience* [37]. That is precisely the fundamental link between game content and player experience. In that regard, all potential content types can elicit player experience. Game content here refers to the game environment, and its impact on player experience can be directly linked to *spatial involvement* and *affective involvement* [3]. But it also includes, as throughout the book, fundamental game-design building blocks such as game mechanics, narrative and reward systems, as well as various other game aspects such as audiovisual settings and camera profiles and effects. In addition complex, social and emotional non-player characters can be used as triggers of desired player experience. In order for agents to elicit meaningful experience and immerse the player they need to engage players in rich and emotional interaction. Towards that purpose they may embed computational models of cognition, behaviour and emotion which are based upon theoretical models such as the OCC [20].

10.3 Modelling player experience

The detection and computational modelling of a user's affective state are core problems in user experience and affective computing research. Detecting and modelling affective states in games can be seen as a special case of this, though in an unusually complex domain. Given the complexity and richness of game-player interaction and the multifaceted nature of player experience, methods that manage to overcome the above challenges and model player experience successfully advance our understanding of human behaviour and emotive reaction with human computer interaction. Player experience modelling (PEM) can thus be viewed as a form of user modelling within games incorporating aspects of behaviour, cognition and affect. PEM involves all three key phases for computational model construction. These are signal processing, feature extraction and feature selection for the model's input; experience annotation for the model's output; and various machine learning and computational intelligence techniques that learn the mapping between the two. Within experience-driven PCG, game content is also represented in the underlying function that characterises player experience.

We can distinguish between *model-based* and *model-free* approaches to player experience modelling [37] as well as potential hybrids between them. The difference is whether the computational model is based on or structured by a theoretical framework. A completely model-based approach relies solely on a theoretical framework that maps game context and player responses to experience. In contrast, a completely model-free approach assumes there is an unknown function between modalities of user input, game content and experience that may be discovered by a machine-learning algorithm (or a statistical model) that does not assume anything about the structure of this function. The space between a completely model-based and a completely model-free approach can be viewed as a continuum along which any PEM approach might be placed. The rest of this section presents the key elements of both model-based and model-free approaches and discusses the core components of a learned computational model (i.e. model input, model output and common modelling methods).

10.3.1 Model input and feature extraction

The PEM's input can be of three main types: a) player behavioural responses to game content as gathered from **gameplay** data (i.e. behavioural data); b) **objective** data collected as player experience manifestations to game content stimuli such as physiology and body movements; and c) the **game context** which comprises any type of game content viewed, played through, and/or created [37, 35, 36].

Given the multifaceted nature of player experience, the input of a PEM usually consists of complex spatio-temporal patterns found in user inputs, sometimes sampled from multiple modalities. These signals need to be processed and relevant data features need to be extracted to feed the model. Relevant features, however, are hard to find within such signals and the ad-hoc design of statistical features often undermines the performance of PEM. There are several available methods within feature extraction (such as principal component analysis and Fischer projection) and feature selection (such as sequential forward selection and genetic-search-based selection) that are applicable to the problem. Recently techniques such as *sequence mining* [15] for feature extraction and *deep learning* [13] for feature combination have shown potential to construct meaningful features for PEM. These methods have been able to fuse data from multiple sources across several player inputs and between player input and game content. In particular, deep learning offers powerful pattern recognition capacities which can detect the most distinct patterns across multiple signals, and provides complex spatio-temporal data attributes that complement standard ad-hoc feature extraction [13]. Sequence mining, on the other hand, identifies the most frequent sequences of events across user input modalities and game context which could be relevant as features for any PEM attempt [15].

In the rest of this section, we will look in more detail at these three types of input to PEM: gameplay input, objective input, and game context input.

10.3.1.1 Gameplay input

The key motivation behind the use of behavioural (gameplay-based) player input is that player actions and real-time preferences are linked to player experience as games affect the player's cognitive processing patterns, cognitive focus and emotional state. Essentially, you express the contents of your mind through gameplay. Arguably it is possible to infer a player's current experience state by analysing patterns of the interaction and associating player experience with game context variables [4, 8]. The models built on this user input type rely on detailed attributes from the player's behaviour which are extracted from player behavioural responses during the interaction with game content stimuli. Such attributes, also named *game metrics*, are statistical spatio-temporal features of game interaction [6] which are usually mapped to levels of cognitive states such as attention, challenge and engagement [23]. In general, both generic measures—such as the level of player performance and the time spent on a task—as well as game-specific measures—such as the items picked and used—are relevant for the gameplay-based PEM.

10.3.1.2 Objective input

The variety of available content types within a game can act as elicitors for complex and multifaceted player experience patterns. Such patterns of experience may, in turn, cause changes in the player's physiology, be reflected in the player's facial expression, posture and speech, and alter the player's attention and focus level. Monitoring such bodily alterations can assist in recognising and synthesising predictors of player experience. The objective approach to PEM assumes access to multiple modalities of player input which manifest aspects of player experience.

Thus, the impact of game content on a number of real-time recordings of the player may be investigated. Physiology offers the primary medium for detecting a player's experience via objective measures [33]: signals obtained from electrocardiography (ECG) [34], photoplethysmography [34, 28], galvanic skin response (GSR) [9], respiration [28], electroencephalography (EEG) [18] and electromyography (among others) are commonly used for the detection of player experience given the recent advancements in sensor technology and physiology-based game interfacing [33]. In addition to physiology the player's bodily expressions may be tracked at different levels of detail and real-time cognitive or affective responses to game content may be inferred. The core assumption of such input modalities is that particular bodily expressions are linked to basic emotions and cognitive processes [2]. Motion tracking may include body posture [22], facial expression and head pose [23].

Beyond the non-verbal cues discussed above there is also room for verbal cue investigation within games. In general, social signals derived from human verbal communication can potentially be used within social games that allow player-to-player interaction (direct or indirect). Such signals challenge the principles of individual player experience modelling but are expected to open the horizon and augment the potential of the experience-driven PCG framework.

10.3.1.3 Game context input

In addition to gameplay and objective data, the context of the game—e.g. the game content experienced, played, or created—is a necessary input for PEM. Game context is the real-time parameterised state of the game which could extend beyond the game content. Without the game context input, player experience models run the risk of inferring erroneous player experience states. For example, an increase in galvanic skin response (GSR) can be linked to a set of dissimilar high-arousal affective states such as frustration and excitement. Thus, the cause of GSR increase (e.g. due to a player's death in a gap between platforms, or alternatively, due to a game level completion) needs to be fused within the GSR signal and embedded in the model. Context-free modelling (while important and desired) has not been investigated to the degree that we can identify generic and context-independent content patterns, features and attributes across games and players. A few recent studies, however, such as that of Martinez et al. [14], attempt to investigate context-independent physiological features that can capture player experience across multiple game genres.

10.3.2 Model output: Experience annotation

The output of a player experience model is provided through an experience annotation process which can either be based on first-person reports (self-reports) or on reports expressed indirectly by experts or external observers [37]. The model's output is, therefore, linked to a fundamental research question within player experience and affective computing: what is the ground truth of player experience and how to annotate it? To address this question a number of approaches have been proposed. The most direct way to annotate player experience is to ask the players themselves about their experience, and build a model based on these annotations. Subjective annotation can be based on either players' free response during play or on forced data retrieved through questionnaires. Alternatively, experts or external observers may annotate the playing experience in a similar fashion. Third-person player experience annotation entails the identification of particular user (cognitive, affective, behavioural) states by user experience and game design experts.

Annotations (either forced self-reports or third-person) can be classified as *rating* (scalar), *class*, or *preference* (ranking) data. With ratings, annotators are asked to answer questionnaire items given in a rating/scaling form—such as the Game Experience Questionnaire [11] or the Geneva Emotion Wheel [1]—which labels user states with a scalar value (or a vector of values). In a class-based format, subjects are asked to pick a user state from a particular representation which is usually a simple boolean question (Was that game level frustrating or not? Is this a sad facial expression?). In the preference annotation format [29], annotators are asked to compare a playing experience in two or more variants/sessions of the game (Was that level more engaging that this level? Which facial expression looks happier?). Recent comparative studies have argued that rating approaches have disadvantages compared to ranking questionnaire schemes [32, 16], such as increased order-of-play and inconsistency effects [30] and lower inter-rater agreement [17, 31].

10.3.3 Modelling approaches

The approach used to construct models of player experience heavily relies on the modelling approach followed (model-based vs. model-free) and the annotation scheme adopted. With the model-based approach, components of the model and any parameters that describe them are constructed in an ad-hoc manner and, sometimes, tested for validity on a trial-and-error basis. No machine learning or sophisticated computational tools are required for these approaches. One could envisage optimising the parameter space to yield more accurate models; that, however, would require empirical studies that bring the approach closer to a model-free perspective.

Model-free approaches, on the other hand, are dependent on the annotation scheme and, in turn, the type of model output available. If data recorded includes either a scalar representation (e.g. via ratings) or classes of annotated labels of user states any of a large number of machine learning (regression and classification) algorithms can be used to build affective models. Available methods include artificial neural networks, Bayesian networks, decision trees, support vector machines and standard linear regression. Alternatively, if experience is annotated in a ranked format, standard supervised-learning techniques are inapplicable, as the problem becomes one of preference learning [7]. Neuro-evolutionary preference learning [29] and rank-based support vector machines [12], along with simpler methods such as

Fig. 10.1: Player responses to losing in IMB. Adapted from [23]

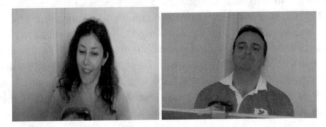

Fig. 10.2: Player responses to winning in IMB. Adapted from [23]

linear discriminant analysis [28], are some of the available approaches for learning preferences.

The ultimate goal of constructing models of player experience is to use these models as measures of content quality and, consequently, to produce affective, cognitive, and behavioural interaction in games and generate personalised or player-adapted content. Quantitative models of player experience can be used to capture player-game interaction and the impact of game content on player experience.

10.4 Example: *Super Mario Bros.*

The work of Shaker et al. [25, 23, 24] on modelling and personalising player experience in *Infinite Mario Bros.* (IMB) [21]—a public-domain clone of *Super Mario Bros.* [19]—gives a complete example of applying the experience-driven PCG approach. First, they build models of player experience based on information collected from the interaction between the player and the game. Different types of features capturing different aspects of player behaviour are considered: *subjective* self-reports of player experience; *objective* measures of player experience collected by extracting information about head movements from video-recorded gameplay sessions; and *gameplay* features collected by logging players' actions in the game. Figures 10.1, 10.2, and 10.3 show examples of objective video data correlated with in-game events: players' reactions when losing, winning, and encountering hard situations, respectively.

Fig. 10.3: Player responses to hard situations in IMB. Adapted from [23]

Table 10.1: The different types of representations of content and gameplay features in [25]

Feature	Description
▬	Flat platform
$(◎)(◎,◎)$	A sequence of three coins
$(R^▸, R^{▸⇑})(♣)$	Moving then jumping in the right direction when encountering an enemy
$(◻, ◻)(◻)$	A gap followed by a decrease in platform height
$(⇑^▸)(S)(▸)$	Jumping to the right followed by standing still then moving right
t_{right}	Time spent moving right
n_{jump}	Total number of jumps
n_{coin}	Total number of coins
k_{stomp}	Number of enemies killed by stomping
N_e	Total number of enemies
B	Total number of blocks

The choice of feature representation is vitally important since it allows different dimensions of player experience to be captured. Furthermore, the choice of content representation defines the search space that can be explored and affects the efficiency of the content-creation method. To accommodate this, the different sets of features collected are represented as frequencies describing the number of occurrences of various events or the accumulated time spent doing a certain activity (such as the number of killings of a certain type of enemies or the total amount of time spent jumping). Features are also represented as sequences capturing the spatial and temporal order of events and allowing the discovery of temporal patterns [25]. Table 10.1 presents example features from each representation.

Based on the features collected, a modelling approach is followed in an attempt to approximate the unknown function between game content, players' behaviour and how players experience the game. The player experience models are developed on different types and representations of features allowing a thorough analysis of the player–content relationship.

The following sections describe the approach followed to model player experience and the methodology proposed to tailor content generation for particular players, using the constructed models as measures of content quality.

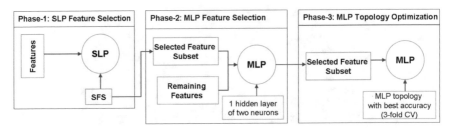

Fig. 10.4: The three-phase player experience modelling approach of [25]

10.4.1 Player experience modelling

When constructing player experience models, the place to start is identifying relevant features of game content and player behaviour that affect player experience. This can be done by recording gameplay sessions and extracting features as indicators of players' affect, performance, and playing characteristics. Given the large size of the feature set that could be extracted, feature selection then becomes a critical step.

In this example, the input space consists of the features extracted from gameplay sessions. Feature selection is done by using sequential forward selection (SFS), a particular feature-selection approach (of many). Candidate features are evaluated by having neuroevolutionary preference learning train simple single-layer perceptrons (SLPs) and multi-layer perceptrons (MLPs) to predict emotional states, and choosing the features that best predict the states [25]. This yields a different subset of features for predicting each reported emotional state.

The underlying function between gameplay, content features, and reported player experience is complex and cannot be easily captured using the simple neuroevolution model used in the feature-selection step. Therefore, once all features that contribute to accurate simple neural network models are found, an optimisation step is run to build larger networks with more complex structures. This is carried out by gradually increasing the complexity of the networks by adding hidden nodes and layers while monitoring the models' performance. Figure 10.4 presents an overview of the process.

Following this approach, models with high accuracies were constructed for predicting players' reports of engagement, frustration and challenge from different subsets of features from different modalities. The models constructed were also of varying topologies and prediction accuracies.

10.4.2 Grammar-based personalised level generator

In Chapter 5, we described how grammatical evolution (GE) can be used to evolve content for IMB. GE employs a design grammar to specify the structure of possible level designs. The grammar is used by GE to transform the phenotype into a level structure by specifying the types and properties of the different game elements that will be presented in the final level design. The fitness function used in that chapter scored designs based on the number of elements presented and their placement properties.

It is possible to use player experience measurements as a component of the fitness function for grammatical evolution as well. This allows us to evolve personalised content. The content is ranked according to the experience it evokes for a specific player and the content generator searches the resulting space for content that maximises particular aspects of player experience. The fitness value assigned for each individual in the population (a level design) in the evolutionary process is the output of the player experience model, which is the predicted value of an emotional state. The PEM's output is calculated by computing the values of the model's inputs; this includes the values of the content features which are directly calculated for each level design generated by GE and the values of the gameplay features estimated from the player's behavioural style while playing a test level.

The search for the best content features that optimise a particular state is guided by the model's prediction of the player experience states, with higher fitness given to individuals that are predicted to be more engaging, frustrating, or challenging for a particular player.

10.4.2.1 Online personalised content generation

Personalisation can be done online. While the level is being played, the playing style is recorded and then used by GE to evaluate each individual design generated. Each individual is given a fitness according to the recorded player behaviour and the values of its content features. The best individual found by GE is then visualised for the player to play.

It is assumed that the player's playing style is largely maintained during consecutive game sessions and thus his playing characteristics in a previous level provide a reliable estimator of his gameplay behaviour in the next level. To compensate for the effect of learning while playing a series of levels, the adaptation mechanism only considers the recent playing style, i.e. the one which the player exhibited in the most recent level. Thus, in order to effectively study the behaviour of the adaptation mechanism, it is important to monitor this behaviour over time. For this purpose, AI agents with varying playing characteristics have been employed to test the adaptation mechanism since this requires the player to playtest a large number of levels. Figure 10.5 presents the best levels evolved to optimise player experience of challenge for two AI agents with different playing styles. The levels clearly exhibit different structures; a slightly more challenging level was evolved for the second agent, with more gaps and enemies than the one generated for the first agent.

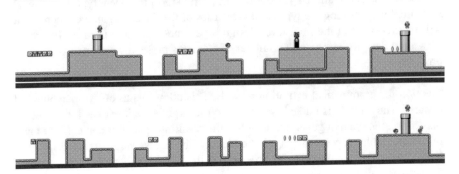

Fig. 10.5: The best levels evolved to maximise predicted challenge for two AI agents. Adapted from [26]

10.5 Lab exercise: Generate personalised levels for *Super Mario Bros.*

In this lab session, you will generate levels personalised for a specific player using the *InfiniTux* software. This is the same software interface used in Chapter 3, but this time the focus is on customising content to a specific playing style.

In order to facilitate meaningful detection of player experience and to allow you to develop player experience models, you will be given a dataset of 597 instances containing several statistical gameplay and content features collected from hundreds of players playing the game. The data contains information about several aspects of players' behaviour captured through features representing the frequencies of performing specific actions such as killing an enemy or jumping and the time spent doing certain behaviour such as moving right or jumping. Your task is to use this data to build a player-experience model using a machine learning or a data-mining technique of your choice. The models you build can then be used to recognise the gameplaying style of a new player.

After you build the models and successfully detect player experience, you should implement a method to adjust game content to changes of player experience in the game. You can adopt well-known concepts of player experience such as *fun*, *challenge*, *difficulty* or *frustration* and adjust the game content according to the aspect you would like your player to experience.

10.6 Summary

This chapter covered the experience-driven perspective for generating personalised game content. The rich and diverse content of games is viewed as a building block to be put together in a way that elicits unique player experiences. The experience-driven PCG framework [37] defines a generic and effective approach for optimising player experience via the adaptation of the experienced content.

To successfully adapt game content one needs to fulfill a set of requirements: the game should be tailored to individual players' experience-response patterns; the game adaptation should be fast, yet not necessarily noticeable; and the experience-based interaction should be rich in terms of game context, adjustable game elements and player input. The experience-driven PCG framework satisfies these conditions via the efficient generation of game content that is driven by models of player experience. The experience-driven PCG framework offers a holistic realization of affective interaction as it elicits emotion through variant game content types, integrates game content into computational models of user affect, and uses game content to adapt the experience.

References

1. Bänziger, T., Tran, V., Scherer, K.R.: The Geneva Emotion Wheel: A tool for the verbal report of emotional reactions. In: Proceedings of the 2005 Conference of the International Society for Research on Emotion (2005)
2. Bianchi-Berthouze, N., Isbister, K.: Emotion and body-based games: Overview and opportunities. In: K. Karpouzis, G.N. Yannakakis (eds.) Emotion in Games: Theory and Praxis. Springer (2016)
3. Calleja, G.: In-Game: From Immersion to Incorporation. MIT Press (2011)
4. Conati, C.: Probabilistic assessment of user's emotions in educational games. Applied Artificial Intelligence **16**(7-8), 555–575 (2002)
5. Csikszentmihalyi, M.: Flow: The Psychology of Optimal Experience. Harper & Row (1990)
6. Drachen, A., Thurau, C., Togelius, J., Yannakakis, G.N., Bauckhage, C.: Game data mining. In: M. Seif El-Nasr, A. Drachen, A. Canossa (eds.) Game Analytics, pp. 205–253. Springer (2013)
7. Fürnkranz, J., Hüllermeier, E. (eds.): Preference Learning. Springer (2011)
8. Gratch, J., Marsella, S.: A domain-independent framework for modeling emotion. Cognitive Systems Research **5**(4), 269–306 (2004)
9. Holmgård, C., Yannakakis, G.N., Karstoft, K.I., Andersen, H.S.: Stress detection for PTSD via the StartleMart game. In: Proceedings of the 5th International Conference on Affective Computing and Intelligent Interaction, pp. 523–528 (2013)
10. Hunicke, R., Chapman, V.: AI for dynamic difficulty adjustment in games. In: Proceedings of the AAAI Workshop on Challenges in Game Artificial Intelligence, pp. 91–96 (2004)
11. IJsselsteijn, W., de Kort, Y., Poels, K., Jurgelionis, A., Bellotti, F.: Characterising and measuring user experiences in digital games. In: Proceedings of the 2007 Conference on Advances in Computer Entertainment Technology (2007)
12. Joachims, T.: Optimizing search engines using clickthrough data. In: Proceedings of the 8th International Conference on Knowledge Discovery and Data Mining, pp. 133–142 (2002)
13. Martínez, H.P., Bengio, Y., Yannakakis, G.N.: Learning deep physiological models of affect. IEEE Computational Intelligence Magazine **8**(2), 20–33 (2013)
14. Martínez, H.P., Garbarino, M., Yannakakis, G.N.: Generic physiological features as predictors of player experience. In: Proceedings of the 4th International Conference on Affective Computing and Intelligent Interaction, pp. 267–276 (2011)
15. Martínez, H.P., Yannakakis, G.N.: Mining multimodal sequential patterns: A case study on affect detection. In: Proceedings of the 13th International Conference on Multimodal Interfaces, pp. 3–10 (2011)

16. Martínez, H.P., Yannakakis, G.N., Hallam, J.: Don't classify ratings of affect; rank them! IEEE Transactions on Affective Computing **5**(3), 314–326 (2014)
17. Metallinou, A., Narayanan, S.: Annotation and processing of continuous emotional attributes: Challenges and opportunities. In: Proceedings of the IEEE Conference on Automatic Face and Gesture Recognition (2013)
18. Nijholt, A.: BCI for games: A 'state of the art' survey. In: Proceedings of the International Conference on Entertainment Computing, pp. 225–228 (2008)
19. Nintendo: (1985). Super Mario Bros., Nintendo
20. Ortony, A., Clore, G., Collins, A.: The Cognitive Structure of Emotions. Cambridge University Press (1990)
21. Persson, M.: Infinite Mario Bros. URL http://www.mojang.com/notch/mario/
22. Savva, N., Scarinzi, A., Berthouze, N.: Continuous recognition of player's affective body expression as dynamic quality of aesthetic experience. IEEE Transactions on Computational Intelligence and AI in Games **4**(3), 199–212 (2012)
23. Shaker, N., Asteriadis, S., Karpouzis, K., Yannakakis, G.N.: Fusing visual and behavioral cues for modeling user experience in games. IEEE Transactions on Cybernetics **43**(6), 1519–1531 (2013)
24. Shaker, N., Togelius, J., Yannakakis, G.N.: Towards automatic personalized content generation for platform games. In: Proceedings of the Artificial Intelligence and Interactive Digital Entertainment Conference, pp. 63–68 (2010)
25. Shaker, N., Yannakakis, G., Togelius, J.: Crowdsourcing the aesthetics of platform games. IEEE Transactions on Computational Intelligence and AI in Games **5**(3), 276–290 (2013)
26. Shaker, N., Yannakakis, G.N., Togelius, J., Nicolau, M., O'Neill, M.: Evolving personalized content for Super Mario Bros using grammatical evolution. In: Proceedings of the Artificial Intelligence and Interactive Digital Entertainment Conference, pp. 75–80 (2012)
27. Sweetser, P., Wyeth, P.: Gameflow: A model for evaluating player enjoyment in games. ACM Computers in Entertainment **3**(3) (2005)
28. Tognetti, S., Garbarino, M., Bonarini, A., Matteucci, M.: Modeling enjoyment preference from physiological responses in a car racing game. In: Proceedings of the IEEE Symposium on Computational Intelligence and Games, pp. 321–328 (2010)
29. Yannakakis, G.N.: Preference learning for affective modeling. In: Proceedings of the 3rd International Conference on Affective Computing and Intelligent Interaction (2009)
30. Yannakakis, G.N., Hallam, J.: Ranking vs. preference: A comparative study of self-reporting. In: Proceedings of the International Conference on Affective Computing and Intelligent Interaction, pp. 437–446 (2011)
31. Yannakakis, G.N., Martínez, H.P.: Grounding truth via ordinal annotation. In: Proceedings of the 6th International Conference on Affective Computing and Intelligent Interaction, pp. 574–580 (2015)
32. Yannakakis, G.N., Martínez, H.P.: Ratings are overrated! Frontiers in ICT **2**, 13 (2015)
33. Yannakakis, G.N., Martínez, H.P., Garbarino, M.: Psychophysiology in games. In: K. Karpouzis, G.N. Yannakakis (eds.) Emotion in Games: Theory and Praxis. Springer (2016)
34. Yannakakis, G.N., Martínez, H.P., Jhala, A.: Towards affective camera control in games. User Modeling and User-Adapted Interaction **20**(4), 313–340 (2010)
35. Yannakakis, G.N., Paiva, A.: Emotion in games. In: R.A. Calvo, S. D'Mello, J. Gratch, A. Kappas (eds.) Handbook of Affective Computing. Oxford University Press (2013)
36. Yannakakis, G.N., Spronck, P., Loiacono, D., Andre, E.: Player modeling. In: Dagstuhl Seminar on Artificial and Computational Intelligence in Games, pp. 45–59 (2013)
37. Yannakakis, G.N., Togelius, J.: Experience-driven procedural content generation. IEEE Transactions on Affective Computing **2**(3), 147–161 (2011)

Chapter 11
Mixed-initiative content creation

Antonios Liapis, Gillian Smith, and Noor Shaker

Abstract Algorithms can generate game content, but so can humans. And while PCG algorithms can generate some kinds of game content remarkably well and extremely quickly, some other types (and aspects) of game content are still best made by humans. Can we combine the advantages of procedural generation and human creation somehow? This chapter discusses mixed-initiative systems for PCG, where both humans and software have agency and co-create content. A small taxonomy is presented of different ways in which humans and algorithms can collaborate, and then three mixed-initiative PCG systems are discussed in some detail: Tanagra, Sentient Sketchbook, and Ropossum.

11.1 Taking a step back from automation

Many PCG methods discussed so far in this book have focused on fully automated content generation. *Mixed-initiative* procedural content generation covers a broad range of generators, algorithms, and tools which share one common trait: they require human input in order to be of any use. While most generators require some initial setup, whether it's as little as a human pressing "generate", or providing configuration and constraints on the output, mixed-initiative PCG automates only part of the process, requiring significantly more human input during the generation process than other forms of PCG.

As the phrase suggests, both a human creator and a computational creator "take the initiative" in mixed-initiative PCG systems. However, there is a sliding scale on the type and impact of each of these creators' initiative. For instance, one can argue that a human novelist using a text editor on their computer is a mixed-initiative process, with the human user providing most of the initiative but the text editor facilitating their process (spell-checking, word counting or choosing when to end a line). At the other extreme, the map generator in *Civilization V* (Firaxis 2014) is a mixed-initiative process, since the user provides a number of desired properties of

(a) Computer-aided design: Humans have the idea, the computer supports their creative process

(b) Interactive evolution: The computer creates content, humans guide it to create content they prefer

Fig. 11.1: Two types of mixed-initiative design

the map. This chapter will focus on less extreme cases, however, where both human and computer have some significant impact on the sort of content generated.

It is naive to expect that the human creator and the computational creator always have equal say in the creative process:

- In some cases, the human creator has an idea for a design, requiring the computer to allow for an easy and intuitive way to realize this idea. Closer to a word processor or to Photoshop, such content generators facilitate the human in their creative task, often providing an elaborate user interface. The computer's initiative is realized as it evaluates the human design, testing whether it breaks any design constraints and presenting alternatives to the human designer. Generators where the creativity stems from human initiative, as seen in Figure 11.1a, will be discussed in Section 11.2.
- In other cases, the computer can autonomously generate content but lacks the ability to judge the quality of what it creates. When evaluating generated content is subjective, unknown in advance, or too daunting to formulate mathematically, generators can request human users to act as judges and guide the generative processes towards content that these users deem better. The most common method for accomplishing this task is interactive evolution, as seen in Figure 11.1b, and discussed in Section 11.3. In interactive evolution the computer has the creative initiative while the human acts as an advisor, trying to steer the generator towards their own goals. In most cases, human users don't have direct control over the generated artifacts; selecting their favourites does not specify how the computer will interpret and accommodate their choice.

11.2 A very short design-tool history

To understand mixed-initiative PCG systems, as well as to gain inspiration for future systems, it is important to also understand several older systems on which current work builds. There are three main threads of work that we'll look at in this section: mixed-initiative interaction, computer-aided design (CAD), and creativity support tools. Today's research in game-design and mixed-initiative PCG tools has been influenced by the ways each of these three areas of work frames the idea of joint human–computer creation [1, 18, 28], and the systems we'll talk about in the chapter all take inspiration from at least one of them.

11.2.1 Mixed-initiative interaction

In 1960, J.C.R. Licklider [24] laid out his dream of the future of computing: man–computer symbiosis. Licklider was the first to suggest that the operator of a computer take on any role other than that of the puppetmaster—he envisioned that one day the computer would have a more symbiotic relationship with the human operator. Licklider described a flaw of existing interactive computer systems: "In the man-machine systems of the past, the human operator supplied the initiative, the direction, the integration, and the criterion."

Notice the use of the term "initiative" to refer to how the human interacts with the computer, and the implication that the future of man-computer symbiosis therefore involves the computer being able to share initiative with its human user.

The term "mixed-initiative" was first used by Jaime Carbonell to describe his computer-aided instruction system, called SCHOLAR [3]. SCHOLAR is a text-based instructional system that largely consists of the computer asking quiz-style questions of the student using the system; the mixed-initiative component of the system allows the student to ask questions of the computer as well. Carbonell argued that there were two particularly important and related aspects of a mixed-initiative system: context and relevancy. Maintaining context involves ensuring that the computer can only ask questions that are contextually relevant to the discussion thus far, ensuring that large sways in conversation do not occur. Relevancy involves only answering questions with relevant information, rather than all of the information known about the topic.

It can be helpful to think about the sharing of initiative in mixed-initiative interaction in terms of a conversation. Imagine, for example, two human colleagues having a conversation in the workplace:

Kevin: "Do you have time to chat about the tutorial levels for the game?"
Sarah: "Yes, let's do that now! I think we need to work together to re-design the first level. Do you—."
Kevin: "Yeah, I agree, players aren't understanding how to use the powerups. I was thinking we should make the tutorial text bigger and have it linger on the screen for longer."
Sarah: "Well, information I got from the user study session two days ago implied that players weren't reading the text at all. I'm not sure if making the text bigger will help."
Kevin: "I think it will help."
          pause
Kevin: "It's easy to implement, at least."

Sarah: "Okay, how about you try that, and I'll work on a new idea I have for
    having the companion character show you how to use them."
Kevin: "Great! Let's meet again next week to see how it worked."

There are several ways in which Kevin and Sarah are sharing initiative in this
conversation. Novick and Sutton [30] describe several components of initiative:

1. Task initiative: deciding what the topic of the conversation will be, and what
   problem needs to be solved. In our example, Kevin takes the task initiative, by
   bringing up the topic of altering the tutorial levels, and by introducing the problem that, specifically, players don't understand how to use the powerups.
2. Speaker initiative: determining when each actor will speak. Mixed initiative is
   often characterized as a form of turn-taking interaction, where one actor speaks
   while the other waits, and vice versa. Our example conversation mostly follows
   a turn-taking model, but deviates in two major areas: a) Kevin interrupts Sarah's
   comments because he thinks he already knows what she will say, and b) Kevin
   later speaks twice in a row, in an effort to move the conversation along.
3. Outcome initiative: deciding how the problem introduced should be solved,
   sometimes involving allocating tasks to participants in the conversation. For this
   example, Sarah takes the outcome initiative, determining which tasks she and
   Kevin should perform as a result of the conversation.

The majority of mixed-initiative PCG systems focus entirely on the second kind
of initiative: speaker initiative. They involve the computer being able to provide
support during the design process, an activity that design researcher Donald Schön
has described as a reflective conversation with the medium [32] (more on this in the
next section). However, they all explicitly give the human designer sole responsibility for determining what the topic of the design conversation will be and how to
solve the problem; all mixed-initiative PCG systems made thus far have prioritized
human control over the generated content.

11.2.2 *Computer-aided design and creativity support*

Doug Engelbart, an early pioneer of computing, posited that computers might augment human intellect. He envisioned a future in which computers were capable of
"increasing the capability of a man to approach a complex problem situation, to
gain comprehension to suit his particular needs, and to derive solutions to problems" [10]. Engelbart argued that all technology can serve this purpose. His de-
augmentation experiment, in which he wrote the same text using a typewriter, a
normal pen, and a pen with a brick attached to it, showed the influence that the
technology has on the ways that we write and communicate.

A peer of Engelbart's, Ivan Sutherland, created the Sketchpad system in 1963
[42]. This was the first system to offer computational support for designers; it was
also the first example of an object-oriented system (though it did not involve programming). Sketchpad allowed designers to specify constraints on the designs they

were drawing; for example, it was possible to draw the general topology of an item such as a bolt. The user could then place constraints on the edges of the bolt to force them to be perpendicular to each other. The system was object-oriented in that individual sketches could be imported into others to produce entire diagrams and drawings; if the original sketch was altered, that change would propagate to all diagrams that imported the sketch. The idea of letting users create through adding and removing constraints has carried forward into mixed-initiative tools such as Tanagra and Sketchaworld, described later in this chapter.

A decade after Engelbart and Sutherland's work, Nicholas Negroponte proposed the creation of what he called design amplifiers. Negroponte was particularly interested in how to support non-expert architectural designers, as he was concerned that professional architects often pushed their own agendas without regard for the needs of the occupants [27]. However, homeowners do not have the domain expertise required to design their own home. His vision was for tools that could help the general population in creating their own homes using the computer to support their designs and ensure validity of the design. The idea that human creators should take the forefront and have the majority of control over a design situation is reflected in all mixed-initiative design tools; in general, the computer is never allowed to override a decision made by the human. However, all tools must push an agenda to some extent, though it may not be intentional: the choices that go into how the content generator operates and what kind of content it is capable of creating vastly influences the work that the human designer can create with the system.

More recently, Chaim Gingold has pushed the idea of "magic crayons": software that supports a novice's creativity while also being intuitive, powerful, and expressive [11]. Gingold argues that using design support tools should be as simple and obvious as using a crayon, and allow for instant creativity. Any child who picks up a crayon can quickly and easily grasp how to use it and go on to create several drawings quite rapidly. The "magic" part of the magic crayon comes in the crayon's computational power and expressive potential: the crayon is imbued with computational support that allows a user to create something better than what they would normally create themselves, while still echoing their original design intent.

11.2.3 Requirements, caveats, and open problems for mixed-initiative systems

When designing a mixed-initiative system, there are several main questions to consider. These points are based on the authors' experiences creating their own prototype mixed-initiative tools:

- *Who is your target audience?*

How to design both the underlying technology and the interface for a mixed-initiative system depends wildly upon who the target audience is. A tool for pro-

fessional designers might look considerably different from a tool intended for game
players who have no design experience.

- *What novel and useful editing operations can be incorporated?*

A mixed-initiative environment offers the opportunity for more sophisticated level-
editing operations than merely altering content as one could do in a non-AI-
supported tool. The way the generation algorithm works might prioritize certain
aspects of the design. For example, Tanagra's underlying generator used rhythm as
a driver for creating levels; thus, it was relatively straightforward to permit users to
interact with that underlying structure to be able to directly manipulate level pacing.

- *How can the method for control over content be balanced?*

Mixed-initiative content generators can involve both direct and indirect manipula-
tion of the content being created. For example, the tool will typically support a user
directly drawing in aspects of the content (e.g. level geometry), but also allow the
computer to take over and make new suggestions for the generator. How to balance
these forms of control can be challenging (especially when the human and computer
conflict, see next point). Should the computer be allowed to make new suggestions
whenever it wants, or only when specifically requested? How much of the content
should be directly manipulable?

- *How to resolve conflicts that arise due to the human stating conflicting desires?*

In situations where both human and computer are editing content simultaneously,
editing conflict inevitably arises. The majority of mixed-initiative tools follow the
principle that the human has final say over what is produced by the tool. However, when the human user states contradictory desires, the system must decide how
to handle the situation. Should it simply provide an error message? Should it randomly choose which desire is more important for the human? Should it generate
several plausible answers and then ask the human to choose which solution is most
reasonable?

   More generally, the issue is: how can the computer infer human design intent via
an interface where the human simply interacts with the content itself.

- *How expressive is the system?*

All content that a human can produce using a mixed-initiative PCG system must be
possible for the computer to generate on its own. Thus it is vital for the system to
be expressive enough to offer a meaningful set of choices to the human user. More
information about expressivity evaluation is in Chapter 12.

- *Can the computer explain itself?*

It is difficult for a human and computer to engage in a design collaboration if neither
is able to explain itself to the other. In particular, a human designer may become
frustrated or confused if the computer consistently acts as though it is not following
the model that the human designer has in her head for how the system should work.
The computer should appear intelligent (even if the choices it is making do not

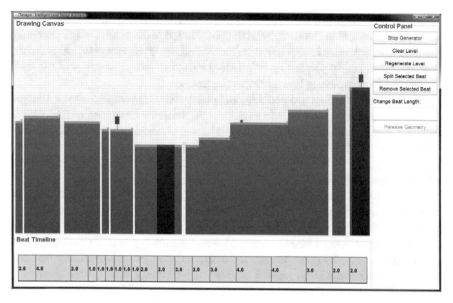

Fig. 11.2: Tanagra, an intelligent level design tool. The level is created in the large area at the upper left. Below is a beat timeline, where the pacing of the level can be manipulated. On the right are buttons for editing the level. Adapted from [40]

involve a sophisticated AI system), and ideally should be able to explain its actions to the human. Being able to communicate at a meta-level about the design tasks and outcomes has not been well explored in mixed-initiative PCG work thus far.

11.2.4 Examples of CAD tools for games

Now we'll go through several examples of computer-aided design tools that both allow interaction and feedback for the human designer and introduce some initiative taken by the computational designer (in varying degrees and in different forms).

11.2.4.1 Tanagra

Tanagra is a mixed-initiative tool for level design, allowing a human and a computer to work together to produce a level for a 2D platformer [40]. An underlying, reactive level generator ensures that all levels created in the environment are playable, and provides the ability for a human designer to rapidly view many different levels that meet their specifications. The human designer can iteratively refine the level by placing and moving level geometry, as well as through directly manipulating the pacing of the level. Tanagra's underlying level generator is capable of producing many different variations on a level more rapidly than human designers, whose strengths instead lie in creativity and the ability to judge the quality of the generated content. The generator is able to guarantee that all the levels it creates are playable, thus refocusing early playtesting effort from checking that all sections of the level are reachable to exploring how to create fun levels.

A combination of reactive planning and constraint programming allows Tanagra to respond to designer changes in real time. A Behavior Language (ABL) [25] is used for reactive planning, and Choco [45] for numerical constraint solving. Reactive planning allows for the expression of generator behaviours, such as placing patterns of geometry or altering the pacing of the level, which can be interleaved with a human designer's actions.

The version of Tanagra displayed in Figure 11.2 incorporates (1) the concept of a user "pinning" geometry in place by adding numerical positioning constraints, (2) the system attempting to minimise the number of required positioning changes (including never being allowed to move pinned geometry), and (3) direct changes to level pacing by adding, removing, and altering the length of beats. Later versions of Tanagra altered the UI to make it clearer what geometry was "pinned" and what was not. The latest version of Tanagra also added the idea of geometry preference toggles, allowing designers an additional layer of control over the system by letting them state whether or not particular geometry patterns are preferred or disliked on a per-beat basis.

11.2.4.2 Sentient Sketchbook

Sentient Sketchbook is a computer-aided design tool which assists a human designer in creating game levels, such as maps for strategy games [22] (shown in Figure 11.3) or dungeons for roguelike games [23]. Sentient Sketchbook uses the notion of map sketch as a minimal abstraction of a full game level; this abstraction limits user fatigue while creating new levels and reduces the computational effort of automatically evaluating such sketches, but is rendered into a high-resolution map before use. Like popular CAD tools, Sentient Sketchbook supports the human creator by automatically testing maps for playability constraints, by calculating and displaying navigable paths, by evaluating the map on gameplay properties and by converting the coarse map sketch into a playable level.

The innovation of Sentient Sketchbook is the real-time generation and presentation of alternatives to the user's sketch. These alternatives are evolved from an initial population seeded by the user's sketch, and thus a certain degree of map integrity is maintained with the user's designs. The shown suggestions are guaranteed to be playable (i.e. have all vital components such as bases and resources for strategy games connected with passable paths) via the use of constrained evolutionary optimisation with two populations [16]. The suggestions are either evolved to maximise one of the predefined objective functions inspired by popular game design patterns such as balance and exploration [23], or towards divergence from the user's current sketch through feasible-infeasible novelty search [21].

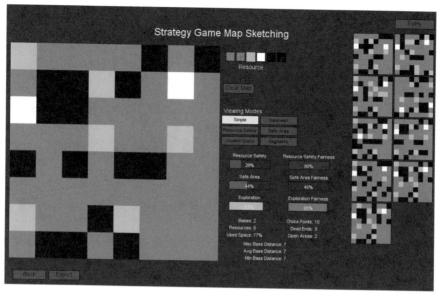

Fig. 11.3: The user interface of Sentient Sketchbook. A human designer edits their sketch (left) and a generator, acting as the artificial designer, creates map suggestions in response (right). Adapted from [22]

11.2.4.3 Ropossum

Ropossum is an authoring tool for the generation and testing of levels of the physics-based puzzle game, *Cut the Rope* [36]. Ropossum integrates many features: (1) automatic design of complete solvable content, (2) incorporation of designer's input through the creation of complete or partial designs, (3) automatic check for playability, and (4) optimisation of a given design based on playability.

Ropossum consists of two main modules: an evolutionary framework for procedural content generation [34] and a physics-based playability module to solve given designs [35]. The second module is used both for evolving playable content and for playtesting levels designed by humans. The parameters of the evolutionary system and the AI agent are optimised so that the system can respond to the user's inputs within a reasonable amount of time. Grammatical evolution (GE) is used to evolve the content. The level structure is defined in a design grammar (DG) which defines the positions and properties of different game components and permits an easy to read and manipulate format by game designers [34]. First-order logic is used to encode the game state as facts specifying the game components and their properties (such as position, speed, and moving direction) [35]. The relationships between the components are represented as rules used to infer the possible next actions.

The two methods for evolving game design and assessing whether the design is playable are combined in a framework to evolve playable content. An initial level design, according to the design grammar, is generated or created by the designer (see

Fig. 11.4: One of the interfaces of Ropossum. The components highlighted are the ones constrained by the designer, and therefore will not be changed when evolving complete playable levels. Adapted from [36]

Figure 11.4) and encoded as facts that can be used by the AI reasoning agent. Given the game state, the agent infers the next best action(s) to perform. The actions are then sent to the physics simulator, which performs the actions according to a given priority and updates the game state accordingly. The new game state is sent to the agent to infer the next action. If the sequence of actions does not lead to winning the level, the system backtracks. A state tree is generated that represents the actions and states explored. For each action performed, a node in the tree is generated and the tree is explored in a depth-first approach. The size of the explored branches in the solution tree is reduced by assigning priorities to the actions and employing domain knowledge encoded in reasoning agent's rules to infer the best action to perform.

11.2.4.4 Sketchaworld

Sketchaworld is an interactive tool created to enable a non-specialist user to easily and efficiently create a complete 3D virtual world by integrating different procedural techniques [39]. Sketchaworld integrates many features: (1) it facilitates easy interaction with designers, who can specify procedural modelling operations and directly visualize their effects, (2) it builds 3D worlds by fitting all features with their surroundings and (3) it supports iterative modelling.

The tool allows user interactions in two main modes: *landscape mode* and *feature mode*. The first mode consists of the user input which is a 2D layout map of the virtual world formatted in a colouring grid that includes information about elevations and soil materials painted with a brush. In the feature mode, users add more specific land features such as cities and rivers (see Figure 11.5).

Fig. 11.5: A screenshot from Sketchaworld. The user sketches as input a mountain, river, and forest, and corresponding 3D terrain is generated. Adapted from [39]

While users sketch on the 2D grid, the effects of their modification are directly visualized in the 3D virtual world. This requires blending the features added with their surroundings. To this end, whenever a new feature is created, an automatic local adaptation step is performed to ensure smoothness and correctness. This includes for example removing trees on a generated road's path or adding a road to connect a generated bridge.

11.3 Interactive evolution

As its name suggests, interactive evolutionary computation (IEC) is a variant of evolutionary computation where human input is used to evaluate content. As artificial evolution hinges on the notion of survival of the fittest, in interactive evolution a human user essentially selects which individuals create offspring and which individuals die. According to Takagi [44], interactive evolution allows human users to evaluate individuals based on their subjective preferences (their own *psychological space*) while the computer searches for this human-specified global optimum in the genotype space (*feature parameter space*); as such, the collaboration between human and computer makes IEC a mixed-initiative approach. In interactive evolution a human user can evaluate artifacts by assigning to each a numerical value (proportionate to their preference for this artifact), by ordering artifacts in order of preference, or by simply selecting one or more artifacts that they would like to see more of. With more control to the human user, the artifacts in the next generation may match users' desires better; the user's cognitive effort may also increase, however, which results in *user fatigue*, which is covered in Section 11.3.1

Interactive evolution is often used in domains where designing a fitness function is a difficult task; for instance, the criteria for selection could be a subjective measure of beauty as in evolutionary art, a deceptive problem where a naive quantifiable measure may be more harmful than helpful, or in cases where mathematically defining a measure of optimality is as challenging as the optimisation task itself. Since it allows for a subjective evaluation of beauty, IEC has often been used to create 2D visual artifacts based on L-systems [26], mathematical expressions [37], neural networks [33] or other methods. Using interactive evolution in art is often motivated by a general interest in artificial life, as is the case with Dawkins' Biomorph [7]. In evolutionary art, human users often evaluate not the phenotypes but outputs specified by the phenotypes, in cases where the phenotypes are image filters or shaders [9]. Apart from 2D visual artifacts, IEC has also been used in generating 3D art [6], animated movies [38], typography [31], and graphic design [8]. Evolutionary music has also used IEC to generate the rhythm of percussion parts [46], jazz melodies [2], and accompaniments to human-authored music scores [14], among others. Outside evolutionary art and music, IEC has been used for industrial design [12], image database retrieval [5], human-like robot motion [29], and many others. The survey by Takagi [44] provides a thorough, if somewhat dated, overview of IEC applications.

11.3.1 User fatigue and methods of combating it

Since interactive evolution is entirely reliant on human input to drive its search processes, its largest weakness is the effect of human fatigue in human-computer interaction. Human fatigue becomes an issue when the users are required to perform a large number of content selections, when feedback from the system is slow, when the users are simultaneously presented with a large amount of content on-screen, or when users are required to provide very specific input. All of these factors contribute to the *cognitive overload* of the user, and several solutions have been proposed to counteract each of these factors.

As human users are often overburdened by the simultaneous presentation of information on-screen, user fatigue can be limited by an interactive evolutionary system that shows only a subset of the entire population. There are a number of techniques for selecting which individuals to show, although they all introduce biases from the tool's designers. An intuitive criterion is to avoid showing individuals which all users would consider unwanted. Deciding which individuals are unwanted is sometimes straightforward; for instance, musical tracks containing only silence or 3D meshes with disconnected triangles. However, such methods often only prune the edges of the search space and are still not guaranteed to show wanted content. Another technique is to show only individuals with the highest fitness; since fitness in interactive evolution is largely derived from user choices, this is likely to result in individuals which are very similar—if not identical—to individuals shown previously, which is more likely to increase fatigue due to perceived stagnation.

User fatigue is often induced when the requirement of a large number of selections becomes time-consuming and cumbersome. As already mentioned, fewer individuals than the entire population can be shown to the user; in a similar vein, not every generation of individuals needs to be shown to the user, instead showing individuals every 5 or 10 generations. In order to accomplish such a behaviour, the fitness of unseen content must be somehow predicted based on users' choices among seen content. One way to accomplish such a prediction is via distance-based approaches, i.e. by comparing an individual that hasn't been presented to the user with those individuals that were presented to the user: the fitness of this unseen individual can be proportional to the user-specified fitness of the closest seen individual while inversely proportional to their distance [15]. Such a technique essentially clusters all individuals in the population around the few presented individuals; this permits the use of a population larger than the number of shown individuals as well as an offline evolutionary sprint with no human input. Depending on the number of seen individuals and the expressiveness of the algorithm's representation, however, a number of strong assumptions are made—the most important of which pertains to the measure of distance used. In order to avoid biasing the search by these assumptions, most evolutionary sprints are only for a few generations before new human feedback is required.

Another solution to the extraneous choices required of IEC systems' users is to crowdsource the selection process among a large group of individuals. Some form of online database is likely necessary towards that end, allowing users to start evolving content previously evolved by another user. A good example of this method is PicBreeder [33], which evolves images created by compositional pattern-producing networks (CPPNs). Since evolution progressively increases the size of CPPNs due to the Neuroevolution of Augmenting Topologies algorithm [41], the patterns of the images they create become more complex and inspiring with large networks. This, however, requires extensive evolution via manual labor, which is expected to induce significant fatigue on a single user. For that reason, the PicBreeder website allows users to start evolution "from scratch", with a small CPPN able to create simple patterns such as circles or gradients, or instead load images evolved by previous users and evolve them further. Since such images are explicitly saved by past users because they are visually interesting, the user starts from a "good" area of the genotype space and is more likely to have meaningful variations than if they were starting from scratch and had to explore a large area of the search space which contains non-interesting images.

Another factor of user fatigue is the slow feedback of evolutionary systems; since artificial evolution is rarely a fast process, especially with large populations, the user may have to sit through long periods of inaction before the next set of content is presented. In order to alleviate that, interactive evolution addresses it by several shortcuts to speed up convergence of the algorithm. This is often accomplished by limiting the population size to 10 or 20 individuals, or by allowing the user to interfere directly in the search process by showing a visualization of the search space and letting them designate an estimated global optimum [43].

To reduce the cognitive load of evaluations, a common solution is to limit the number of rating levels, either to a common five-star rating scale, or even to only two: the user either likes the content or doesn't. Another option is to use rankings [17], i.e. the user is presented with two options and chooses the one they prefer, without having to explicitly specify that e.g. one is rated three stars while the other is five-star content.

11.3.2 Examples of interactive evolution for games

As highly interactive experiences themselves, games are ideal for interactive evolution, since the user's preferences can be inferred from what they do in the game. Instead of an explicit selection process, selection masquerades behind in-game activities such as shooting, trading, or staying alive. Done properly, interactive evolution in games can bypass to a large extent the issue of user fatigue. However, the mapping between player actions and player preference is often not straightforward; for instance, do humans prefer to survive in a game level for a long time, or do they like to be challenged and be constantly firing their weapons? Depending on the choice of metric (in this example, survival time or shots fired), different content may be favoured. Therefore, gameplay-based evaluations may include more biases on the part of the programmer than traditional interactive evolution, which tries to make no assumptions.

11.3.2.1 Galactic Arms Race

Galactic Arms Race [13], shown in Figure 11.6, is one of the more successful examples of a game using interactive evolution. The procedurally generated weapon projectiles, which are the main focus of this space-shooter game, are evolved interactively using gameplay data. The number of times a weapon is fired is considered a revealed user preference; the assumption is that players who don't like a weapon will not use it as much as others. Weapon projectiles, represented as particles, are evolved via neuroevolution of augmenting topologies (NEAT); the velocity and colour of each particle is defined as the output of a CPPN, with the input being the current position and distance from the firing spaceship.[1] Newly evolved weapons are dropped as rewards for destroying enemy bases; the player can pick them up, and use them or switch among three weapons at any given time. *Galactic Arms Race* can be also played by many players; in multiplayer play, the algorithm uses the firing rates of all players when determining which weapons to evolve.

---

[1] NEAT and CPPNs are discussed in detail in Chapter 9.

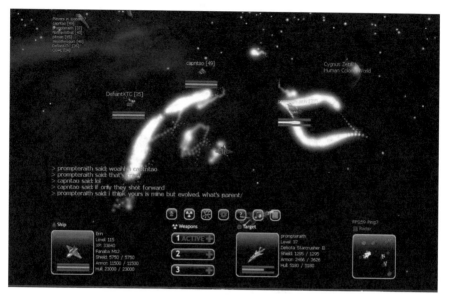

Fig. 11.6: *Galactic Arms Race* with multiple players using different weapons. Adapted from [13]

11.3.2.2 TORCS track generation

A more traditional form of interactive evolution, in which a user directly states preferences, was used to generate tracks for a car racing game [4]. The system uses The Open Racing Car Simulator (TORCS)[2] and allows user interaction through a web browser where users can view populations of race tracks and evaluate them (see Figure 11.7). This web front-end then communicates with an evolutionary backend. Race tracks are represented in the engine as a list of segments which can be either straight or turning. In the evolution process, a set of control points and Bézier curves are used to connect the points and ensure smoothness.

Different variations of interactive evolution are used to evaluate the generated tracks. In *single-user mode*, human subjects were asked to play 10 generations of 20 evolved tracks each and evaluate them using two scoring interfaces: like/dislike and rating from 1 to 5 stars. The feedback provided by users about each track is the fitness used for evolution. In *multi-user mode*, the same population of 20 individuals is played and evaluated by five human subjects. The fitness given to each track in the population is the average score received from all users. The feedback provided by users showed improvements in the quality of the tracks and an increase in their interestingness.

---

[2] http://torcs.sourceforge.net/

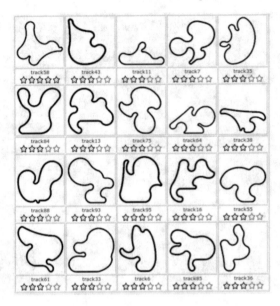

Fig. 11.7: The TORCS track generator visualizes tracks, and asks the player to rank them. Adapted from [4]

11.3.2.3 Spaceship generation

Liapis et al.'s [20] work on spaceship generation is an example of fitness prediction for the purpose of speeding up and enhancing the convergence of interactive evolution. They generate spaceship hulls and their weapon and thruster topologies in order to match a user's visual taste as well as conform to a number of constraints aimed at playability and game balance [19]. The 2D shapes representing the spaceship hulls are encoded as pattern-producing networks (CPPNs) and evolved in two populations using the feasible-infeasible two-population approach (FI-2pop) [16]. One population contains spaceships which fail ad-hoc constraints pertaining to rendering, physics simulation, and game balance, and individuals in this population are optimised towards minimising their distance to feasibility. Removing such spaceships from the population shown to the user reduces the chances of unwanted content and reduces user fatigue.

The second population contains feasible spaceships, which are optimised according to ten fitness dimensions pertaining to common attributes of visual taste such as symmetry, weight distribution, simplicity, and size. These fitness dimensions are aggregated into a weighted sum which is used as the feasible population's fitness function. The weights in this quality approximation are adjusted according to a user's selection among a set of presented spaceships (see Figure 11.8). This adaptive aesthetic model aims to enhance the visual patterns behind the user's selection and minimise visual patterns of unselected content, thus generating a completely new

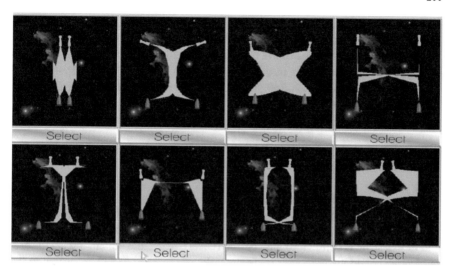

Fig. 11.8: In this evolutionary spaceship generator, the user is presented a set of spaceships from the feasible population, and selects their favourite. Adapted from [20]

set of spaceships which more accurately match the user's tastes. A small number of user selections allows the system to recognize the user's preference, reducing fatigue.

The two-step adaptation system, where (1) the user implicitly adjusts their preference model through content selection and (2) the preference model affects the patterns of generated content, is intended to make for a flexible tool both for personalizing game content to an end-user's visual taste and also for inspiring a designer's creative task with content guaranteed to be playable, novel, and conforming to the intended visual style.

11.4 Exercise

1. Choose one of the tools described in this chapter. Perform a design task similar to that which is supported by the tool without any computational support. Reflect upon this process: What was easy and what was hard? What did you wish the computer could do to help? What do you feel the computer would not be able to assist with? If the tool is available for download, try to perform the same design task using the AI-supported tool. What were some of the key differences in your experience as a designer?
2. Create a requirements analysis document and mock-up architecture diagram for a mixed-initiative design tool that operates in a domain of your choice. Make sure to consider: (a) Who is your audience? (b) What, specifically, is your domain?

(c) What is the PCG system capable of creating? (d) What is the mixed-initiative conversational model the system will follow?
3. Create a paper prototype of the tool you designed in exercise two. Test the prototype with someone else in the class, with you acting as the "AI system" and your partner acting as the designer. Be careful to only act according to how the AI system itself would be able to act.

11.5 Summary

Mixed-initiative systems are systems where both humans and computers can "take the initiative," and both contribute to the creative process. The degree to which each party takes the initiative and contributes varies between different systems. At one end of this scale is computer-aided design (CAD), where the human directs the creative process and the computer performs tasks when asked to and according to the specifications of the user. At the other end is interactive evolution, where the computer proposes new artifacts and the user is purely reactive, providing feedback on the computer's suggestions. Both of these approaches have a rich history in games: computer-aided design in many game design tools that include elements of content generation, and interactive evolution in games such as *Galactic Arms Race*. "True" mixed-initiative interaction, or at least the idea of such systems, has a long history within human-computer interaction and artificial intelligence. Within game content generation, there are so far just a few attempts to realize this vision. Tanagra is a platformer level-generation system that uses constraint satisfaction to complete levels sketched by humans, and regenerates parts of levels to ensure playability as the levels are edited by humans. Sentient Sketchbook assists humans in designing strategy game levels, providing feedback on a number of quality metrics and autonomously suggesting modifications of levels. Ropossum is a level editor for *Cut the Rope*, which can test the playability of levels and automatically regenerate parts of levels to ensure playability as the level is being edited.

References

1. Almeida, M.S.O., da Silva, F.S.C.: A systematic review of game design methods and tools. In: Proceedings of the International Conference on Entertainment Computing, pp. 17–29 (2013)
2. Biles, J.A.: GenJam: A genetic algorithm for generating jazz solos. In: Proceedings of the International Computer Music Conference, pp. 131–137 (1994)
3. Carbonell, J.R.: Mixed-initiative man-computer instructional dialogues. Ph.D. thesis, Massachusetts Institute of Technology (1970)
4. Cardamone, L., Loiacono, D., Lanzi, P.L.: Interactive evolution for the procedural generation of tracks in a high-end racing game. In: Proceedings of the Conference on Genetic and Evolutionary Computation, pp. 395–402 (2011)
5. Cho, S.B., Lee, J.Y.: Emotional image retrieval with interactive evolutionary computation. In: Advances in Soft Computing, pp. 57–66. Springer (1999)

6. Clune, J., Lipson, H.: Evolving three-dimensional objects with a generative encoding inspired by developmental biology. In: Proceedings of the European Conference on Artificial Life, pp. 144–148 (2011)

7. Dawkins, R.: The Blind Watchmaker. W. W. Norton & Company (1986)

8. Dipaola, S., Carlson, K., McCaig, G., Salevati, S., Sorenson, N.: Adaptation of an autonomous creative evolutionary system for real-world design application based on creative cognition. In: Proceedings of the International Conference on Computational Creativity (2013)

9. Ebner, M., Reinhardt, M., Albert, J.: Evolution of vertex and pixel shaders. In: Genetic Programming, *Lecture Notes in Computer Science*, vol. 3447, pp. 261–270. Springer (2005)

10. Engelbart, D.C.: Augmenting human intellect: A conceptual framework. Air Force Office of Scientific Research, AFOSR-3233 (1962)

11. Gingold, C.: Miniature gardens & magic crayons: Games, spaces, & worlds. Master's thesis, Georgia Institute of Technology (2003)

12. Graf, J.: Interactive evolutionary algorithms in design. In: Proceedings of the International Conference on Artificial Neural Nets and Genetic Algorithms, pp. 227–230 (1995)

13. Hastings, E.J., Guha, R.K., Stanley, K.O.: Automatic content generation in the Galactic Arms Race video game. IEEE Transactions on Computational Intelligence and AI in Games **1**(4), 245–263 (2009)

14. Hoover, A.K., Szerlip, P.A., Stanley, K.O.: Interactively evolving harmonies through functional scaffolding. In: Proceedings of the Conference on Genetic and Evolutionary Computation (2011)

15. Hsu, F.C., Chen, J.S.: A study on multi criteria decision making model: Interactive genetic algorithms approach. In: IEEE International Conference on Systems, Man, and Cybernetics, vol. 3, pp. 634–639 (1999)

16. Kimbrough, S.O., Koehler, G.J., Lu, M., Wood, D.H.: On a feasible-infeasible two-population (FI-2Pop) genetic algorithm for constrained optimization: Distance tracing and no free lunch. European Journal of Operational Research **190**(2), 310–327 (2008)

17. Liapis, A., Martínez, H.P., Togelius, J., Yannakakis, G.N.: Adaptive game level creation through rank-based interactive evolution. In: Proceedings of the IEEE Conference on Computational Intelligence and Games (2013)

18. Liapis, A., Yannakakis, G.N., Alexopoulos, C., Lopes, P.: Can computers foster human users' creativity? Theory and praxis of mixed-initiative co-creativity. Digital Culture & Education **8**(2), 136–153 (2016)

19. Liapis, A., Yannakakis, G.N., Togelius, J.: Neuroevolutionary constrained optimization for content creation. In: Proceedings of the IEEE Conference on Computational Intelligence and Games, pp. 71–78 (2011)

20. Liapis, A., Yannakakis, G.N., Togelius, J.: Adapting models of visual aesthetics for personalized content creation. IEEE Transactions on Computational Intelligence and AI in Games **4**(3), 213–228 (2012)

21. Liapis, A., Yannakakis, G.N., Togelius, J.: Enhancements to constrained novelty search: Two-population novelty search for generating game content. In: Proceedings of the Conference on Genetic and Evolutionary Computation (2013)

22. Liapis, A., Yannakakis, G.N., Togelius, J.: Sentient Sketchbook: Computer-aided game level authoring. In: Proceedings of the 8th International Conference on the Foundations of Digital Games, pp. 213–220 (2013)

23. Liapis, A., Yannakakis, G.N., Togelius, J.: Towards a generic method of evaluating game levels. In: Proceedings of the Artificial Intelligence for Interactive Digital Entertainment Conference, pp. 30–36 (2013)

24. Licklider, J.C.R.: Man-computer symbiosis. IRE Transactions on Human Factors in Electronics **1**(1), 4–11 (1960)

25. Mateas, M., Stern, A.: A behavior language for story-based believable agents. IEEE Intelligent Systems **17**(4), 39–47 (2002)

26. McCormack, J.: Interactive evolution of L-system grammars for computer graphics modelling. In: Complex Systems: From Biology to Computation, pp. 118–130. ISO Press (1993)

27. Negroponte, N.: Soft Architecture Machines. MIT Press (1975)
28. Nelson, M.J., Mateas, M.: A requirements analysis for videogame design support tools. In: Proceedings of the 4th International Conference on the Foundations of Digital Games, pp. 137–144 (2009)
29. Nojima, Y., Kojima, F., Kubota, N.: Trajectory generation for human-friendly behavior of partner robot using fuzzy evaluating interactive genetic algorithm. In: Proceedings of the IEEE International Symposium on Computational Intelligence in Robotics and Automation, pp. 114–116 (2003)
30. Novick, D., Sutton, S.: What is mixed-initiative interaction? In: Proceedings of the AAAI Spring Symposium on Computational Models for Mixed Initiative Interaction (1997)
31. Schmitz, M.: genoTyp, an experiment about genetic typography. In: Proceedings of Generative Art (2004)
32. Schön, D.A.: Designing as reflective conversation with the materials of a design situation. Research in Engineering Design 3(3), 131–147 (1992)
33. Secretan, J., Beato, N., D'Ambrosio, D.B., Rodriguez, A., Campbell, A., Stanley, K.O.: Picbreeder: Evolving pictures collaboratively online. In: CHI '08: Proceeding of the 26th SIGCHI Conference on Human factors in Computing Systems, pp. 1759–1768 (2008)
34. Shaker, M., Sarhan, M.H., Naameh, O.A., Shaker, N., Togelius, J.: Automatic generation and analysis of physics-based puzzle games. In: Proceedings of the IEEE Conference on Computational Intelligence and Games, pp. 1–8 (2013)
35. Shaker, M., Shaker, N., Togelius, J.: Evolving playable content for Cut the Rope through a simulation-based approach. In: Proceedings of the Conference on Artificial Intelligence and Interactive Digital Entertainment, pp. 72–78 (2013)
36. Shaker, M., Shaker, N., Togelius, J.: Ropossum: An authoring tool for designing, optimizing and solving Cut the Rope levels. In: Proceedings of the Conference on Artificial Intelligence and Interactive Digital Entertainment, pp. 215–216 (2013)
37. Sims, K.: Artificial evolution for computer graphics. In: Proceedings of the 18th Conference on Computer Graphics and Interactive Techniques, SIGGRAPH '91, pp. 319–328 (1991)
38. Sims, K.: Interactive evolution of dynamical systems. In: Towards a Practice of Autonomous Systems: Proceedings of the First European Conference on Artificial Life, pp. 171–178 (1992)
39. Smelik, R.M., Tutenel, T., de Kraker, K.J., Bidarra, R.: Interactive creation of virtual worlds using procedural sketching. In: Proceedings of Eurographics, pp. 29–32 (2010)
40. Smith, G., Whitehead, J., Mateas, M.: Tanagra: Reactive planning and constraint solving for mixed-initiative level design. IEEE Transactions on Computational Intelligence and AI in Games 3(3), 201–215 (2011)
41. Stanley, K.O., Miikkulainen, R.: Evolving neural networks through augmenting topologies. Evolutionary Computation 10(2), 99–127 (2002)
42. Sutherland, I.E.: Sketchpad: A man-machine graphical communication system. In: Proceedings of the Spring Joint Computer Conference, AFIPS '63, pp. 329–346 (1963)
43. Takagi, H.: Active user intervention in an EC search. In: Proceedings of the International Conference on Information Sciences, pp. 995–998 (2000)
44. Takagi, H.: Interactive evolutionary computation: Fusion of the capabilities of EC optimization and human evaluation. Proceedings of the IEEE 89(9), 1275–1296 (2001)
45. The Choco Team: Choco: An open source Java constraint programming library. In: 14th International Conference on Principles and Practice of Constraint Programming (2008)
46. Tokui, N., Iba, H.: Music composition with interactive evolutionary computation. In: International Conference on Generative Art, pp. 219–226 (2000)

Chapter 12
Evaluating content generators

Noor Shaker, Gillian Smith, and Georgios N. Yannakakis

Abstract Evaluating your content generator is a very important task, but difficult to do well. Creating a game content generator in general is much easier than creating a good game content generator—but what is a "good" content generator? That depends very much on what you are trying to create and why. This chapter discusses the importance and the challenges of evaluating content generators, and more generally understanding a generator's strengths and weaknesses and suitability for your goals. In particular, we discuss two different approaches to evaluating content generators: visualizing the expressive range of generators, and using questionnaires to understand the impact of your generator on the player. These methods could broadly be called top-down and bottom-up methods for evaluating generators.

12.1 I created a generator, now what?

The entirety of this book thus far has been focused on how to create procedural content generators, using a variety of techniques and for many different purposes. We hope that, by now, you have gained an appreciation for the strengths and weaknesses of different approaches to PCG, and also the surprises that can come from writing a generative system. We imagine that you also have experienced some of the frustration that can come from debugging a generative system: "is the interesting level I created a fluke, a result of a bug, or a genuine result?"

Creating a generator is one thing; evaluating it is another. Regardless of the method followed, generators are evaluated on their ability to achieve the desired goals of the designer (or the computational designer). This chapter reviews methods for achieving that. Arguably, the generation of any content is trivial; the generation of *valuable* content for the task at hand, on the other hand, is a rather challenging procedure. Further, it is more challenging to generate content that is both *valuable* and *novel*.

What makes the evaluation of content (such as stories, levels, maps, etc.) difficult is the subjective nature of players, their large diversity and, on the other end of the design process, the designer's variant intents, styles, and goals [9]. Furthermore, content quality is affected by algorithmic stochasticity (such as metaheuristic search algorithms) and human stochasticity (such as unpredictable playing behaviour, style, and emotive responses) that affect content quality at large. All these factors are obviously hard to control in an empirical fashion.

In addition to factors that affect content quality, there are constraints (hard or soft ones) put forward by the designers, or imposed by other elements of game content that might conflict with the generated content (e.g. a generated level must be compatible with a puzzle). A PCG algorithm needs to be able to satisfy designer constraints as part of its quality evaluation. We have seen several types of such algorithms in this book, such as the answer-set programming approach in Chapter 8 and the feasible-infeasible two-population genetic algorithm used in Chapter 11. The generated results in these cases satisfy constraints, thereby they have a certain value for the designer (at least if the designer specified the correct constraints!). But *value* has varying degrees of success, and which constraints to choose are not always obvious, and that is where the methods and heuristics discussed in this chapter can help.

PCG can be viewed as a computational creator (either assisted or autonomous). One important aspect that has not been investigated in depth is the aesthetics and creativity of PCG within game design. How creative can an algorithm be? Is it deemed to have appreciation, skill, and imagination [4]? Evaluating creativity of current PCG algorithms, a case can be made that most of them possess skill but not creativity. Does the creator manage to explore novel combinations within a constrained space thereby resulting in *exploratory* game design creativity [1]; or, is on the other hand trying to break existing boundaries and constraints within game design to come up with entirely new designs, demonstrating *transformational* creativity [1]? If used in a mixed-initiative fashion, does it enhance the designer's creativity by boosting the possibility space for her? The appropriateness of evaluation methods for autonomous PCG creation or mixed-initiative co-creation [19] remains largely unexplored within both human and computational creativity research.

Content generators exhibit highly emergent behaviour, making it difficult to understand what the results of a particular generation algorithm might be when designing the system. When making a PCG system, we are also creating a large amount of content for players to experience, thus it is important to be able to evaluate how successful the generator is according to players who interact with the content. The next section highlights a number of factors that make evaluating content generators important.

12.2 Why is evaluation important?

There are several main reasons that we want to be able to evaluate procedural content generation systems:

1. To better understand their capabilities. It is very hard to understand what the capabilities of a content generator are solely by seeing individual instances of their output.
2. To confirm that we can make guarantees about generated content. If there are particular qualities of generated content that we want to be able to produce, it is important to be able to evaluate that those qualities are indeed present.
3. To more easily iterate upon the generator by seeing whether what it is capable of creating matches the programmer's intent. As with any creative endeavor, creating a procedural content generator involves reflection, iteration, and evaluation.
4. To be able to compare content generators to each other, despite different approaches. As the community of people creating procedural content generators continues to grow, it is important to be able to understand how we are making progress in relationship to the current state of the art.

This chapter describes strategies for evaluating content generators, both in terms of their capabilities as generative systems and in performing evaluations of the content that they create. The most important concept to remember when thinking of how to evaluate a generator is the following: make sure that the method you use to evaluate your generator is relevant to what it is you want to investigate and evaluate. If you want to be able to make the claim that your generator produces a wide variety of content, choose a method that explicitly examines qualities of the generator rather than individual pieces of content. If you want to be able to make the claim that players of a game that incorporates your generator find the experience more engaging, then it is more appropriate to evaluate the generator using a method that includes the player.

One of the ultimate goals of evaluating content generators is to check their ability to meet the goals they are intended to achieve while being designed. Looking at individual samples gives a very high-level overview of the capabilities of the generators but one would like for example to examine the frequency with which specific content is generated or the amount of variety in the designs produced by the system. It is therefore important to visualize the space of content covered by a generator. The effects of modifications made to the system can then be easily identified in the visualized content space as long as the dimensions according to which the content is plotted are carefully defined to reflect the goals intended when designing the system.

The remainder of the chapter covers two main approaches for evaluating content: the *top-down* approach using content generation statistics, in particular *expressivity measures* (see Section 12.3), and the *bottom-up* approach which associates content quality with user experience and direct or indirect content annotations (see Section 12.4).

12.3 Top-down evaluation via expressivity measures

A tempting way to evaluate the quality of a content generator is to simply view the content it creates and evaluate the artifacts subjectively and informally. But if a content generator is capable of creating thousands, even millions, of unique levels, it is not feasible to view all of the output to judge whether or not the generator is performing as desired. If you see five levels that are impressive, among 50 that you choose to ignore or re-generate, what does that say about the qualities of the content generator?

To solve this problem, it is possible to evaluate the expressive range of the level generator. *Expressive range* refers to the space of potential levels that the generator is capable of creating, including how biased it is towards creating particular kinds of content in that space [15]. This evaluation is performed by choosing metrics along which the content can be evaluated, and using those metrics as axes to define the space of possible content. A large number of pieces of content are then generated and evaluated according to the defined metrics and plotted in a heatmap. This heatmap can reveal biases in the generator, and comparisons of the heatmap across different sets of input parameters can show how controllable the generator is. Seeing how expressive range shifts according to input changes yields good insights on the controllability and the quality of your generator.

12.3.1 Visualizing expressive range

The expressive range of a content generator can be visualized as an N-dimensional space, where each dimension is a different quality of the generator that can be quantified. This allows us to imagine authoring level generators as creating these spaces of potential levels as a result of the emergent qualities of the system. By adding and removing rules from a rule-based generator, the shape of the generator's expressive range (also referred to as a generative space) can be altered.

For only two dimensions, the generative space can be visualized using a two-dimensional histogram. Higher dimensionality requires more sophisticated visualizations of generative spaces, which have not been deeply explored in PCG. This requires generating a representative sample of the content and ranking it according to the metrics; determining the amount of content to generate can be tricky. While it is simple for some systems to compute the total number of variations that can be generated, others may be able to create infinite variety. One method to ensure an acceptable sample size in the case of infinite content is to generate increasingly large amounts of content and visualize the expressive range, stopping when the graphs begin to look the same as for the previous, smaller amount of content. Expressive range charts are not intended to be perfect, mathematical proofs of variety; rather, they are a visualization that can help the creator of a generative system to understand its behaviour, and potential users of that system to understand its abilities.

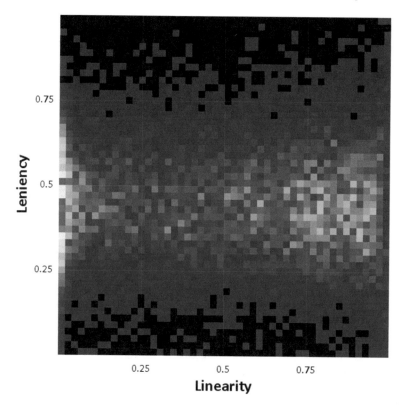

Fig. 12.1: The expressive range of the Launchpad level generator. Adapted from [16]

Figure 12.1 shows the expressive range of the Launchpad level generator [16]. Notice that there is one large hot-spot for creating medium-leniency, low-linearity levels, and another bias towards creating medium-leniency, high-linearity levels (more on these metrics in the next section). Understanding that the system is biased towards these areas forces the designer of the system to ask why such biases exist.

Figure 12.2 presents an alternative method for visualizing the expressive range of one of the content generators for *Infinite Mario Bros.* [13]. The figure shows different distributions of the levels according to three expressive measures defined: linearity, leniency, and density.

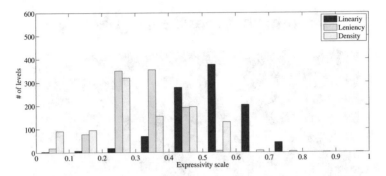

Fig. 12.2: The histograms of the linearity, leniency and density measures for one of the *Infinite Mario Bros.* generators. Adapted from [13]

12.3.2 Choosing appropriate metrics

The metrics used for any content generator are bound to vary based on the domain that content is being generated for. The "linearity" and "leniency" metrics used in the Launchpad generator mentioned above make sense in the context of 2D platforming levels, but perhaps not in the context of weapons for a space shooting game.

The important rule of thumb to remember when choosing metrics for your content generator is the following: strive to choose metrics that are as far as possible from the input parameters to the system. The goal of performing an expressive range evaluation is to understand the emergent properties of the generative system. Choosing a metric that is highly correlated to one that is used as an input parameter (e.g. if your generator accepts "difficulty" as an input and has "difficulty" as an expressive range metric) can only ever provide confirmatory results. If the system is specifically designed to create a particular kind of output, measuring for that output can only show that the algorithm operates as expected; it cannot deliver insight into unexpected behaviour or surprising output.

12.3.3 Understanding controllability

An important consideration in procedural content generation is understanding how well the generator can be controlled to produce different kinds of output, and especially how small changes in rule systems or priorities alter the expressivity of the system. If a designer requests that the system create shorter levels, will it still be capable of producing a broad range of content? Will adding a new rule to the grammar fundamentally alter the qualities of levels that can be produced?

Insight into these issues can be visualized by comparing expressive range graphs for different configurations of input parameters. That can be achieved through vi-

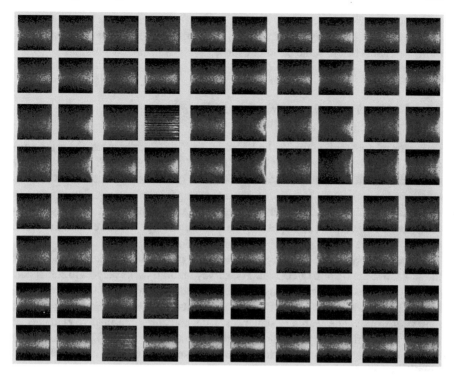

Fig. 12.3: Expressive ranges corresponding to different input parameters. Adapted from [16]

sual comparisons showing the expressive range of the Launchpad level generator when varying its rhythm input parameters (length of segment, pacing of segment, and type of rhythm; see Figure 12.3). Notice that, while several graphs look quite similar to each other, there are notable parameter configurations that lead to drastically different resulting spaces. Gaining insight into the problem is helpful not only after creating a system and wanting to evaluate it, but also during the development process itself as a debugging tool.

12.4 Bottom-up evaluation via players

Complementary to the top-down approaches for the evaluation of content generation, quantitative user studies can be of immense benefit for content quality assurance. The most obvious approach to evaluate the content experienced by players is to *explicitly* ask them about it. A game user study can involve a small number of dedicated players that will play through varying amounts of content or, alternatively, a crowd-sourced approach that can provide sufficient data to machine-learn content evaluation functions (see [14, 8, 3] among others). It is important to note that any bottom-up content quality assessment can be complemented by content annotations of the designers of the game or other experts involved in content creation.

12.4.1 Which questionnaire should I use?

Estimates of content quality can be based on a player's preferences about the content retrieved through questionnaires during or after the gameplay. Content quality can be represented as a class, a scalar (or a vector of numbers)—such as continuous annotation throughout a level—or a relative strength (preference). Content quality can be characterized by a number of dimensions, such as novelty, feasibility, playability, and believability [2], depending on the content type evaluated and the scope of the evaluation.

There are several user protocol schemes and questionnaires one can adopt for the subjective annotation of content. The survey naturally asks players to self-report their experiences about content using directed questions which can vary from simple tick boxes to multiple choice items. Both the questions and the answers provided may vary from single words to sentences. Questionnaires can involve elements of the player experience (e.g. the Game Experience Questionnaire [7]), demographic data, or other factors that might impact the assessment of content quality (e.g. personality traits).

As a general recommendation for self-reporting it is suggested that subjects should be asked to *rank* the content they experience [21, 17]. Our recommendation is based on studies by Yannakakis and Martínez showing limitations when using the more common choice of ratings for subjective assessment, which provide empirical evidence for the superiority of rank-based questionnaire schemes for subjective annotation (in games and beyond) [20, 21, 10]. In particular, rating-based questionnaires generate lower inter-rater agreement and higher levels of inconsistency and order effects, and are dominated by a number of critical biases that make any post-hoc analysis questionable [21, 20, 18, 11].

12.4.2 Ways around the limitations of self-reporting

While self-reporting biases can be minimised by using rank-based questionnaires, not all disadvantages of self-reporting can be avoided this way [21]. Self-reports can be replaced by or triangulated with alternate measures of player experience such as physiological manifestations (of e.g. arousal, interest, and attention) [22] and/or behavioural playing patterns [5] that may map to content quality directly. For instance, a player being stuck at the same point of a map for several minutes might indicate player frustration and bad level design. Further, the rank-based approach can be enhanced if users are given the opportunity to view and annotate their video playthroughs (e.g. as in [6]). Evidently such a survey protocol further eliminates reporting biases associated with memory and cognitive load which are present when users are asked to annotate the content in a post-experience manner.

Even though the data-driven approach for annotating content is user-centric and promotes the assessment of content quality directly by its end users, it has another core limitation which is associated with the potential treatment of subjects as random content evaluators. Thus, ideally, the generator should be coupled with selection mechanisms that will prune the available content (which arguably can be generated in massive amounts automatically) prior to it being presented to the players. On that basis, content can be evaluated via a sequence of logic operations without the need for player behavioural metrics or other input from players [12]. The satisfaction of constraints or logical operators can be coupled with simulations of AI agents that evaluate the quality of generated content (i.e. offline generate-and-test PCG). Further, the quality of content can be evaluated *implicitly* through player preferences during gameplay. It is natural to assume that the more content is selected, used, experienced, or altered during gameplay, the higher its value becomes. Studies in weapon generation, for instance, estimate the weapon's quality implicitly by its use during the game (see Section 11.3.2.1). and without explicitly asking the players about the quality of the weapons or other characteristics. Such an approach however relies heavily on the assumption that content use indicates content quality.

12.5 Summary

This chapter focused on the important but complex task of content evaluation, and discussed various methods to achieve it. In summary, content quality can either be assessed in a top-down fashion via well-designed content statistics (such as expressivity metrics) or in a bottom-up fashion by having players experience the content and assess its quality. Content evaluation via players may rely on both behavioural data and data for objective assessment such as physiology, such as the methods discussed in Chapter 10. This chapter, instead, focused on content evaluation via subjective user data as obtained from questionnaires. Arguably a hybrid approach involving both top-down and bottom-up methods can provide a more holistic approach to game content evaluation, and ultimately to *understanding* what the generator does, and whether it is suitable for the job the designer wants to use it for.

References

1. Boden, M.A.: What is creativity? In: M.A. Boden (ed.) Dimensions of Creativity, pp. 75–117. MIT Press (1994)
2. Camilleri, E., Yannakakis, G.N., Dingli, A.: Platformer level design for player believability. In: Proceedings of the IEEE Conference on Computational Intelligence and Games (2016)

3. Chernova, S., Orkin, J., Breazeal, C.: Crowdsourcing HRI through online multiplayer games. In: AAAI Fall Symposium: Dialog with Robots, pp. 14–19 (2010)
4. Colton, S.: Creativity versus the perception of creativity in computational systems. In: Proceedings of the AAAI Spring Symposium on Creative Intelligent Systems, pp. 14–20 (2008)
5. Drachen, A., Thurau, C., Togelius, J., Yannakakis, G.N., Bauckhage, C.: Game data mining. In: M. Seif El-Nasr, A. Drachen, A. Canossa (eds.) Game Analytics, pp. 205–253. Springer (2013)
6. Holmgård, C., Yannakakis, G.N., Martínez, H.P., Karstoft, K.I.: To rank or to classify? Annotating stress for reliable PTSD profiling. In: Proceedings of the International Conference on Affective Computing and Intelligent Interaction, pp. 719–725 (2015)
7. IJsselsteijn, W., De Kort, Y., Poels, K., Jurgelionis, A., Bellotti, F.: Characterising and measuring user experiences in digital games. In: Proceedings of the ACE 2007 Workshop on Methods for Evaluating Games (2007)
8. Li, B., Lee-Urban, S., Appling, D.S., Riedl, M.O.: Crowdsourcing narrative intelligence. Advances in Cognitive Systems **2**, 25–42 (2012)
9. Liapis, A., Yannakakis, G.N., Togelius, J.: Towards a generic method of evaluating game levels. In: Proceedings of the Artificial Intelligence for Interactive Digital Entertainment Conference, pp. 30–36 (2013)
10. Martínez, H.P., Yannakakis, G.N., Hallam, J.: Don't classify ratings of affect; rank them! IEEE Transactions on Affective Computing **5**(3), 314–326 (2014)
11. Metallinou, A., Narayanan, S.: Annotation and processing of continuous emotional attributes: Challenges and opportunities. In: Proceedings of the IEEE Conference on Automatic Face and Gesture Recognition (2013)
12. Nelson, M.J.: Game metrics without players: Strategies for understanding game artifacts. In: Proceedings of the First AIIDE Workshop on AI in the Game-Design Process, pp. 14–18 (2011)
13. Shaker, N., Nicolau, M., Yannakakis, G.N., Togelius, J., ONeill, M.: Evolving levels for Super Mario Bros. using grammatical evolution. In: Proceedings of the IEEE Conference on Computational Intelligence and Games, pp. 304–311 (2012)
14. Shaker, N., Yannakakis, G., Togelius, J.: Crowd-sourcing the aesthetics of platform games. IEEE Transactions on Computational Intelligence and AI in Games **5**(3), 276–290 (2013)
15. Smith, G., Whitehead, J.: Analyzing the expressive range of a level generator. In: Proceedings of the First Workshop on Procedural Content Generation in Games (2010)
16. Smith, G., Whitehead, J., Mateas, M., Treanor, M., March, J., Cha, M.: Launchpad: A rhythm-based level generator for 2-d platformers. IEEE Transactions on Computational Intelligence and AI in Games **3**(1), 1–16 (2011)
17. Yannakakis, G.N.: Preference learning for affective modeling. In: Proceedings of the International Conference on Affective Computing and Intelligent Interaction (2009)
18. Yannakakis, G.N., Hallam, J.: Ranking vs. preference: A comparative study of self-reporting. In: Proceedings of the International Conference on Affective Computing and Intelligent Interaction, pp. 437–446 (2011)
19. Yannakakis, G.N., Liapis, A., Alexopoulos, C.: Mixed-initiative co-creativity. In: Proceedings of the 9th Conference on the Foundations of Digital Games (2014)
20. Yannakakis, G.N., Martínez, H.P.: Grounding truth via ordinal annotation. In: Proceedings of the International Conference on Affective Computing and Intelligent Interaction, pp. 574–580 (2015)
21. Yannakakis, G.N., Martínez, H.P.: Ratings are overrated! Frontiers in ICT **2**, 13 (2015)
22. Yannakakis, G.N., Martínez, H.P., Garbarino, M.: Psychophysiology in games. In: K. Karpouzis, G.N. Yannakakis (eds.) Emotion in Games: Theory and Praxis. Springer (2016)

Appendix A
Game-designer interviews

To complement the technical content of the chapters, all written by academics (though some of the chapter authors also design and develop games), we performed five interviews with the creators of well-known PCG-heavy games. We selected the interviewees mostly because the games they had been part of had either introduced new interesting PCG techniques, or because they had integrated them in game design in some novel way. The interviews were performed in 2013 and 2014 over email, and are reproduced in their entirety here (except for corrected typos). We asked most of the interviewees the same set of questions, focusing on the role of PCG in game design and the limits of generative methods.

The interviewees are:

- **Andrew Doull**, creator of *UnAngband* and *UnBrogue*, founder of RogueLikeRadio.
- **Ed Key**, creator of *Proteus*.
- **Michael Toy**, co-creator of *Rogue*.
- **Richard Evans**, AI lead programmer on *The Sims 3*, co-creator of *Versu*.
- **Tarn Adams**, creator of *Dwarf Fortress*.

A.1 Andrew Doull

Was there anything you wanted to do in a game you worked on that you could not do because of algorithmic or computational limitations?

My thinking about game design has changed significantly over the years, since writing the original Death of the Level Designer series of articles. One of the key changes—guided a lot by games like Michael Brough's 868-HACK—is that a game should embrace limitations, rather than attempt to design around them. So rather than creating procedural systems which can model everything à la Dwarf Fortress,

© Springer International Publishing Switzerland 2016
N. Shaker et al., *Procedural Content Generation in Games*, Computational
Synthesis and Creative Systems, DOI 10.1007/978-3-319-42716-4

there are real advantages in keeping games as limited as possible in order that the significance of individual procedural elements is emphasized, rather than smoothed over by a melange of inputs. My personal limitations are very much around algorithm implementation rather than design: for instance, while I have a good understanding of what a Voronoi diagram looks like and where it could be used, I'm unlikely to ever successfully reimplement anything but a brute force approach for calculating cell membership.

What new design questions has PCG posed for some game you worked on?

I've written extensively about this—refer to the writing on my blog post on Unangband's dungeon generation and algorithmic monster placement for specific discoveries. Since then, UnBrogue includes very little original procedural content: I mostly plug new values into the well written framework that Brian Walker has developed for Brogue's "machine" rooms. These days I try to steer clear of actually designing procedural systems: my experience is that you can achieve a lot using a very simple set of algorithms, provided you choose your content carefully (see Darius Kazemi's essay on Spelunky's level generation for a great example of this).

What is the most impressive example of procedural content generation you have seen since your own work?

I'd be hard pressed to ignore Miguel Cepero of Voxel Farm, who I'm sure a lot of people you interview will mention. While the above ground system looks great, it was his cave designs that won me over, after being a doubter. What really impresses me though is he's developing all this while being the father of twins—I'm in the same position and I can never find the time...

What do you think of the fact that roguelikes have become a genre of their own? Is PCG in your opinion an essential part of what a roguelike is?

I'm going to quote Edmund McMillen here, since he made the definitive statement on why you should write a roguelike:
    "The roguelike formula is an amazing design plan that isn't used much, mostly because its traditional designs rely on alienatingly complicated user interfaces. Once you crack the roguelike formula, however, it becomes an increasingly beautiful,

deep, and everlasting design that allows you to generate a seemingly dynamic experience for players, so that each time they play your game they're getting a totally new adventure."[1]

PCG is obviously an important part of this process, but it isn't independent from the other roguelike genre features like permadeath.

My hunch is that there are other "design plans" out there that are waiting to be found that will feature PCG—in fact the majority will do, but we don't necessarily know what they look like, or have the maturity of the medium (of PCG) to be able to discover them.

What are your tips for designing games that use PCG?

Keep your algorithms simple and choose your content carefully.

Do you have any interesting stories about PCG failures?

Not personally, since I've taken such a conservative approach to PCG algorithms.

In general, is there anything in a game you think could never be procedurally generated?

The specific quirks of the real world. I'm not saying that PCG can't create something like the real world: I expect the depth required to make a world "completely convincing" is actually more shallow than most PCG "haters" realize—but ultimately, when it comes to simulation, the fact that we exist at a specific time and place in a continuum of choices and random events is something that PCG can only hold a mirror up to. Hand placed design will always be needed if you want to model history—PCG will overtake hand placed design for "fantasy worlds" in the not too distant future.

Why is PCG not used more?

PCG is a language that requires a level of literacy to understand. We're not effectively teaching this language yet, but we're not effectively teaching the language of game design in general either. Also, it is often more expensive than hand placed con-

---

[1] http://www.gamasutra.com/view/feature/182380/

tent, because a PCG algorithm which is only 90% complete can not create anything useful, whereas 90% accurate hand placed content is clearly 9/10ths done. It's hard to describe working with PCG this way, but there's almost a phase change between when a PCG algorithm just creates junk, and when it starts producing beautiful results, and it can be very hard to tune it to reach this state.

What do you see as current directions for PCG that are worth investigating?

There's a lot of interesting stuff happening on the academic side—getting this to percolate over to game development is going to be the real challenge.

A.2 Ed Key

Was there anything you wanted to do in a game you worked on that you could not do because of algorithmic or computational limitations?

At first there was: At one point Proteus was going to be some kind of sandbox RPG with generated towns and quests. Once I started talking to David about music, we reshaped the game as being all about music and exploration, and also at this point started to find and work with the "grain" (as in carving wood) of what we had, shaping what we wanted to do to the medium we were working in.

You can extend this to Proteus being an island rather than an infinitely streaming world. The latter would have been technically much harder but also not really desirable once we decided to focus on a finite space that allowed you to get a little lost but was also bounded and so allowed some familiarity and revisiting of locations.

Do you have any tips for designing games that use procedural content?

Think about framing, structure and pacing. Consider how in the classical example of Rogue, the procedural generation is incredibly simple and designed as a kind of "lumpy canvas" for the authored elements (creature, potions, etc.) to interact. I think the most successful PCG applications understand how the procedural content is framed and given context by authored content, or sometimes by human curation. Think of procedural generation as poetry or music and make use of the player's imagination and faith rather than trying to create results that withstand point-by-point examination.

One strength of PCG is to create "wildness"—either mimicking or evoking nature or in glitch aesthetics. On the other hand, formal disciplines like architecture provide patterns that PG systems can use, but I think you still need something in the "fiction" of the game to make freakish "wrong" results something appealing rather than bugs that break immersion.

Something I'm really keen on in game design is how to create "substances" or things that "feel substantial". I think a lot of this is about establishing scope and language early on and sticking to these as a contract with the player. Of course, you can subvert those expectations, but first you need to establish them.

In general, is there anything in a game you think could never be procedurally generated?

Stuff like human behaviour is always going to be hard. My solution to this is to have the PCG operate at a level of hints and suggestions instead of trying to generate fully detailed characters, behaviour and artefacts. If the player is invested enough to fill in the gaps with their imagination this will be a much richer experience than if they are just given all the details and their mind unengaged and free to pick holes in those.

There's a deeper issue in that "meaning" can never be created by a computer system, in my opinion. "Meaning" arises in the mind of a conscious being and is about how the player reads and interacts with the game. On the other side it's about what you as the architect of the PG system put into it—values, aesthetics, etc. Humanity is paradoxically extremely important in this domain.

What is the most impressive example of procedural content generation you have seen since your own work?

My friend Alex May is doing some beautiful stuff with procedural trees[2]. For something that's released and generating a full gameworld, maybe this PG stuff by Tom Betts[3]. No Man's Sky is also extremely enticing but hard to separate hype and expectation from the actual product at the current time (Jan 2013).

---

[2] http://blog.starboretum.com/

[3] http://www.big-robot.com/tag/sir-you-are-being-hunted/

What do you think of the fact that roguelikes have become a genre of their own? Is PCG in your opinion an essential part of what a roguelike is?

Well, roguelike when I first knew it was permadeath and proc-gen ascii dungeons. Now we have a whole spectrum of roguelike-likes including FTL, Don't Starve, etc. I think the genre already existed but has become broader, whilst at the same time procedural techniques are spreading and growing in all genres from FPSes to interaction fiction. I would say that yes, PCG is essential to a roguelike, but it's always interesting to take that as a challenge. Maybe the great PCG-free roguelike is Dark Souls? No-one calls that a roguelike, and I think it wouldn't work if it was procedurally generated, but it seems to share a lot of the flavour of "punishing exploration". It's interesting to think about how Dark Souls would be *worse* if it was proc-gen. Places in the world would have less resonance, and players wouldn't be able to share stories or advice in the same way.

A.3 Michael Toy

Was there anything you wanted to do in Rogue that you could not do because of algorithmic or computational limitations?

The limited size of programs on the PDP 11/70 (64 kilobytes), kept us from implementing the variety of AI driven monsters that we had imagined in the design phase.

What is the most impressive example of procedural content generation you have seen since your own work?

Have to tip my hat to Dwarf Fortress. The story-telling and emergent properties are marvelous. And the game Moria was probably the closest to what we had imagined doing when we started writing Rogue. I'm sure there are more, I am not an expert in the field, but there's my answer.

What do you think of the fact that roguelikes have become a genre of their own? Is PCG in your opinion an essential part of what a roguelike is?

The word "roguelike" belongs to the community that invented the word, so I don't claim any special authority. However the initial design goal for Rogue was to produce a game that avoided two problems, and the two solutions resulting are often stated in the definition of a roguelike, PCG and permadeath.

The first problem was that having written several text adventures, it eventually (it should have been sooner, we were young and stupid) became clear that it was never going to be fun playing a game where you knew everything. So the quest became to try and make a game where even the creator of the game is involved in a quest for discovery.

Second we wanted to avoid the "Dragon's Lair" problem where winning the game is just running until you die, then backing up and doing something different, repeated endlessly. We allowed saved games so you could stop and go to class or eat, but worked hard to dis-allow people from re-playing from a save point repeatedly, not because we were trying to create permadeath precisely, but because we wanted the in-game consequences to matter. If a player decided to take a small or a large risk, we wanted that risk to be a more real risk than simply the risk that you might have to restore from the save file. This then made the rewards more meaningful also. It wasn't just permadeath, but perma-everything.

I actually see PCG and prevention of reverse time-jumps as being inseparable. If I can save the game, explore a level, restore at the save point and explore the level again only "correctly", the entire point of the PCG is missed.

What are your tips for designing games that use PCG?

I think PCG changes how you think about the world-writing for a game.

One of the surprises for us in writing Rogue is how little PCG it took to create a game which people could play for hundreds of hours. We really barely got working what we thought was the base game, and suddenly it was popular and everywhere, before we got to what we had previously thought was going to be the part which made it interesting.

In a sense a game of Rogue is a collaborative storytelling exercise. You don't know how it is going to end, though you suspect it will be a tragedy. People have imaginations and that can be a huge advantage to game designers. Rogue allowed people to write their own scripts about what was going on, and provided them all the action scenes for their story.

In general, is there anything you think could never be generated?

I always dream of PCG worlds as rich and beautiful as the best hand-modeled worlds. Not because I think the modeling is trivial, or even possible to do, but because I love the feeling of stepping into something that has never been seen before. The problem is that a world which takes your breath away is not just doors and walls, it is cultures and civilizations. Even traditional games rarely invent these things, but just re-skin the ones we already know about. Can you ever generate something like walking through a jungle and discovering ruins in a style that no human has seen before?

A.4 Richard Evans

Would you describe The Sims 3 as doing procedural narrative generation? What about Versu?

It depends, as always, on your criteria.

The Sims games create a broad range of possible permutations of behaviour. Often, the generated behaviour sequence is everyday—but sometimes the behaviour sequence seems to conform to a narrative. The richer the personality model, and the deeper the social simulation, the more likely this is to happen. Certainly, some people did create narratives just by sitting back and watching The Sims 3. For example, Robin Burkinshaw created Alice and Kev: a great blog describing the plight of a couple of homeless Sims. He set up an initial situation (a father-daughter pair, who were both homeless), and then sat back, watching and recording the events as they occurred.

Versu procedurally generates narrative. At the drama-manager level (the level of scenes), there is a moderately rigid story graph of scenes with pre- and post-conditions. But within each particular scene, the individual agents are free to choose their own actions, based on their own desires.

Interactive storytelling originates in its own separate community. As interactive stories develop more generative procedural-storytelling systems, do they become a kind of procedural-content domain?

Interactive stories often have hand-written content at the drama-manager level, but variation and procedurality within the scene. This is a limited form of procedurality existing inside a constrained hand-written framework.

Was there anything you wanted to procedurally generate in a game you worked on that you could not do because of algorithmic or computational limitations?

Yes. One thing I really wanted the computer to do was to generate a social situation that was already half-way through. So if, for example, the player turned up at a bus-stop, there might be an argument between a boy and a girl that was already almost finishing. This ability, to create social situations in media res, is not something that the Versu simulator is able to do.

What is the most impressive example of procedural content generation you have seen?

Ooo I don't know. There are so many recent exciting examples. Procedurally generated platform levels, music composition, puzzle games—it's a very exciting time.

Do you have any interesting stories about procedural-content failures?

The richer the simulation, the more possible causal pathways—and the harder it can be to understand why something is happening. During development of Versu, I had a tricky bug where half way through a murder-mystery, the doctor was being rather rude to my player character. I know that the doctor did not have an abrasive personality, and my character had never done anything rude to the doctor, so it was hard to see why the doctor was behaving this way. It turned out, after much debugging, that the reason was this: at the beginning of the game, my player character had been dismissive to one of the servants who was waiting at the table. The servant had gone back to the kitchen, and had told the others about my rude behaviour. The doctor, being a friend of the servant, had believed the servant's testimony and had formed a negative judgement about my player's character. This sort of example shows how emergence is a double-edged sword: it generates new stories, some of which are not anticipated—but can also make it harder to understand what is happening.

In general, is there anything in a game you think could never be procedurally generated?

In Versu, we generated text from templates (e.g. "[X] look[s] towards [Y obj]"), substituting proper names and pronouns for variables (generating e.g. "Jack looks towards her"). What would be significantly harder—but also significantly more flexible—would be to generate text without templates—using e.g. a phrase-structure grammar and semantic constraints.

A.5 Tarn Adams

Was there anything you wanted to generate in Dwarf Fortress, or another game you worked on that you could not do because of algorithmic or computational limitations?

Most of the algorithms are scalable, so it's really that almost everything needs to be kept smaller than we'd like. Things like time travel are difficult to do in a proper fashion in DF since the amount of data is extreme, and even a small perturbation wouldn't be believable if it didn't have a lot of data to back it up. Fluid dynamics are difficult, and our system is pretty lame due to computational problems (and algorithmic/scientific cluelessness for anything complicated there). All of the conversation AI is very basic and will likely be held back by a lack of good ideas on my part and also my dislike of generating a lot of English sentences. The entire frontier of what we haven't done in DF is made up of our limitations along these lines, when it isn't just time constraints.

What new design questions has PCG posed for some game you worked on?

One of the interesting ones is the matter of presentation. If you generate most of the content in the game, and it doesn't hew to traditional lines, you have to be careful about how you unfurl it to the player. We're just getting started with this consideration now as we start making more generated creatures and plant-life and materials, but the game would become gray mush if we aren't mindful as we move away from pre-defined content. A very simple example is the paragraph description it pops up whenever a forgotten beast attacks your fortress—if the player were attacked without being forced to look at a description, I think it would become quite confusing as the random attacks and other properties of the creature come into play, though there's a lot of wiggle room and different methods that could be tried out. When we allow the game to replace regular wilderness creatures or the standard fantasy races (elf, goblin, etc.) or even the playable race, the roll-out of the random characteristics is going to be front and center... almost tutorial-worthy in some cases.

There's also the matter of the realized map area which has come up a lot— sometimes Dwarf Fortress has pieces of the world loaded up at five different levels of abstraction (or more?), and each of those need to mesh with each other and be chosen so that crucial details aren't lost but also so that memory and speed are under control. This can be difficult to manage, and sometimes we have to make choices that make the game suffer—this would relate back to the first question regarding algorithm scalability I guess. You can often accomplish a lot of what you want to accomplish without doing a perfect job by just scaling back the loaded area a bit, or keeping an abstract version of a larger area loaded (whether that's a map area or some other concept).

What is the most impressive example of procedural content generation you have seen since your own work?

The cities from Subversion, maybe? Although it wasn't fully realized, it seems like it would have been cool. Seeing the game unfold in Drox Operative was neat, although if you view yourself as an RPG camera in a strategy game there, it probably doesn't stand out as an AI example. The complete package was interesting though.

What do you think of the fact that roguelikes have become a genre of their own? Is PCG in your opinion an essential part of what a roguelike is?

I don't know that genres are ever healthy, but it's cool to see more games. I don't have a definition for "roguelikes", and it's a popular subject for argument, but I can't think of a game without map-related PCG that I'd ever casually call a "roguelike" to somebody in conversation. Other people focus more on permanent death, save states, and other features, though, and for all I know Gauntlet is a "roguelike" now.

What are your tips for designing games that use PCG?

For all of these, there's the caveat that rules are meant to be broken after looking at the bigger picture, and also that the tips grew out of mistakes I still make which are evident in my games. So: Don't simulate more detail than you need to get your point across— the elements involved in the PCG should be game elements, atmospheric elements, etc.—if you don't need the molecules bumping around, invisible to the player, try to stay away from chewing up computer resources and programming time putting that in. It's fine to go one level deeper if the "phenotype" that arises from your procedural DNA turns out well, but that's a matter of happy accidents as much as planning, and you'll have difficulty refining your game if you subject yourself to too much chaos theory (or to too much going on that just doesn't affect anything).

Keep in mind what the game is trying to accomplish overall. If you can afford it, don't substitute crappy PCG for a single, better hand-crafted asset (unless there's a really solid counter-balance, say, in replayability, then it is a matter of taste)—at least if you are polishing up your game, since experimentation is crucial at first and you might arrive at something really satisfying. PCG does not automatically increase replayability—a full play through a great pre-defined game is better than a full play through a shoddy random game that you won't touch again. I haven't always been able to do the following in a timely fashion, since it can hamper experimentation, but if parts of your game are moddable, I think it's good to keep your own internal PCG in line with the moddable format (to keep the standards uniform if anything else)—for example, PCG Dwarf Fortress creatures and materials are made by producing a text definition which is interpreted in the same way as pre-defined or modded text definitions.

I also think it is good to try your hand at your own algorithms when possible, since the output will have more character than something recognizable as Perlin noise or a Voronoi diagram etc., though just having anything you can iterate on is probably good enough, and of course existing algorithms form an important part of your PCG skills to build on. Sometimes you can get better results faster just by plowing ahead, though, rather than fishing around for something online. Most things haven't been tried yet, and there's a wide frontier of PCG in games to explore. If you are simulating something that can be related to a real-world process, keep that process in mind when you are trying to correct unacceptable defects in your output—the answer is often in some missing variable or relationship that the real-world analogy makes clear.

Do you have any interesting stories about PCG failures?

They're mostly interesting from the humor angle, since things often go terribly wrong. I'm not sure what would be interesting for the experts or people interested in making better PCG. My process is very iterative, and it's difficult for me to remember discrete instructive moments.

In general, is there anything in a game you think could never be procedurally generated?

You can view everything produced by people throughout history as procedurally generated in a larger context, so I wouldn't leave anything out in general, although there's probably a Gödel-ish proof sitting around that you can't PCG everything from computer algorithms. Some things are certainly more difficult than others.

Prose and conversations and so on can be rough, especially as it relates to AI (since that's just the Turing test more or less), and chaotic behavior that comes from many small parts (like fluids or weather) is probably not possible since you'd need to simulate the molecular behavior properly to hit upon the actual effects (though you could use a "good enough" test like the Turing test, for dynamic behaviors vs. human observation of them). So in any case, the actual limits of PCG probably aren't important yet, and I don't think supposed limitations related to being able to match pre-defined human artwork in terms of emotional impact or symbolic significance or whatever else should deter anybody from exploring what's possible.

Why is PCG not used more?

People are using it now more than I've ever seen, so I'm not sure this one is answerable from my perspective. If current PCG techniques don't measure up, it's prudent to stick with hand-crafted text and graphics and music from a financial and overall quality perspective, certainly, rather than trying to tackle everything with PCG to your satisfaction in the time you have available for your project.

What do you see as current directions for PCG that are worth investigating?

I think people are already jumping into everything, in one way or another, and nothing should be off limits there. I look at PCG as an almost universal candidate for feature implementation (since I don't have useful skills for producing non-PCG material), so I wouldn't close off or prefer any avenue. There's a good game to be made regarding any subject, and PCG can be involved in those.

Printed in the United States
By Bookmasters